# Contents

KT-436-658

## Unit 3 Practical skills in chemistry 1

iv

# AS
# Chemistry

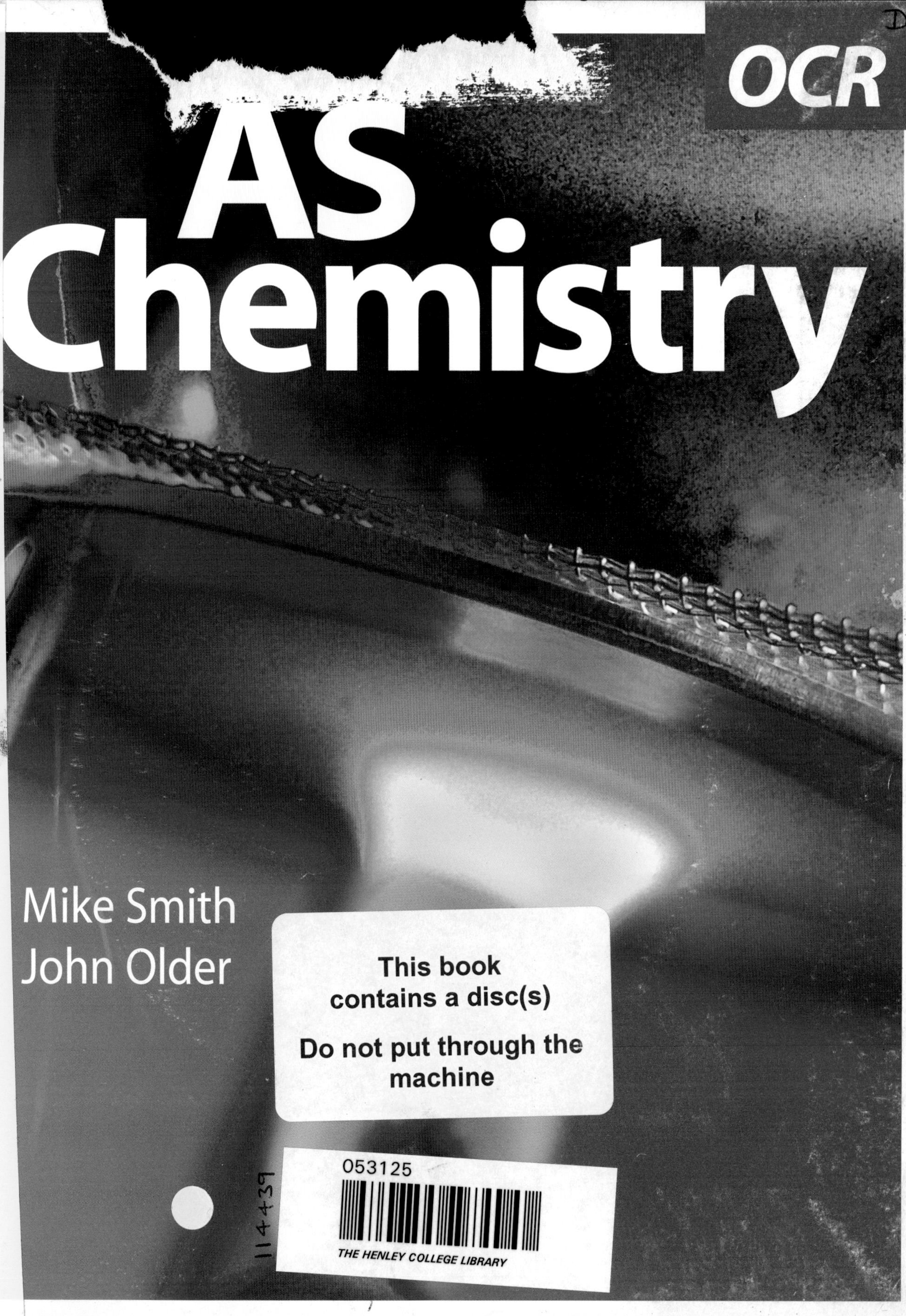

Mike Smith
John Older

OCR

Philip Allan Updates, an imprint of Hodder Education, part of Hachette Livre UK, Market Place, Deddington, Oxfordshire OX15 0SE

*Orders*
Bookpoint Ltd, 130 Milton Park, Abingdon, Oxfordshire, OX14 4SB
tel: 01235 827720    fax: 01235 400454
e-mail: uk.orders@bookpoint.co.uk
Lines are open 9.00 a.m.–5.00 p.m., Monday to Saturday, with a 24-hour message answering service. You can also order through the Philip Allan Updates website: www.philipallan.co.uk

ISBN 978-1-84489-434-5

First printed 2008
Impression number 5 4 3 2
Year 2013 2012 2011 2010 2009 2008

This textbook has been written specifically to support students studying OCR AS Chemistry. The content has been neither approved nor endorsed by OCR and remains the sole responsibility of the authors.

All efforts have been made to trace copyright on items used.

Design by Juha Sorsa
Printed in Italy

Hachette Livre UK's policy is to use papers that are natural, renewable and recyclable products and made from wood grown in sustainable forests. The logging and manufacturing processes are expected to conform to the environmental regulations of the country of origin.

P01322

# Introduction

## About this book

This textbook covers the OCR specification for AS chemistry introduced in 2008. In general, the order of the topics corresponds to that of the specification. However, some reactions of acids and redox reactions are covered in Chapter 8 with the chemistry of groups 2 and 7. Students starting an AS course have varying backgrounds and some may not need to study Chapters 1 and 2 in detail. Care has been taken, however, to ensure that those with more limited experience have the necessary information to make a success of the course. In particular, experience suggests that some students struggle to provide correct formulae for substances and consequently fail to construct balanced equations. This topic is, therefore, covered in detail.

The first seven chapters describe and explain the material of **Unit 1: Atoms, bonds and groups**. The unit is divided into three component modules. The work on atoms, moles and equations is covered in the first five chapters and some of the reactions of acids are also found there. Chapter 6 includes the information and ideas about bonding and structure that are required for the second module. The third module, which includes the periodic table and groups 2 and 7, is covered in Chapter 7. Redox reactions and the specified reactions of acids complete this chapter.

**Unit 2: Chains, energy and resources** is addressed in detail in Chapters 8–13. Chapters 8–11 cover the first two modules and detail the reactions of the organic functional groups, including the knowledge of mass spectrometry and infrared spectroscopy that is required. Chapter 12 covers the module on energy, reaction rates and equilibrium. Environmental concerns and 'green' chemistry are covered in Chapter 13.

**Unit 3: Practical skills in chemistry 1** consists of centre-based practical assessment. Practical work is often ignored in textbooks. Yet there are particular, and sometimes elusive, skills to be acquired through carrying out experiments. Chapter 14 addresses this issue and provides students with the knowledge they need in order to approach the assessments with confidence.

Within the specification, reference is made to 'How science works'. This builds on an important feature of the revised GCSE courses. 'How science works' covers a range of issues — from the historical context in which new ideas and theories were suggested, to the way in which new materials have been developed and the increasing importance of chemistry in combating environmental threats. It emphasises the breadth of knowledge and understanding that is required of a well-educated scientist.

The constant mention of 'How science works' has been avoided in this book, partly because it is an all-inclusive issue and partly because it is more directly relevant to teachers than it is to students. It is, however, likely to affect the context in which an examination question is set, rather than its content.

At the end of each chapter, there is an additional reading section. This is intended to provide further, sometimes challenging, information. The brief biographical profiles of some of the scientists who advanced our knowledge of chemistry may also serve to provide a human touch to the work that they did.

The additional reading section has another purpose, which is to address what is expressed in the specification as 'stretch and challenge'. The inclusion of an A* grade in the examination results at A2 means that the most able students will need a broader knowledge and a deeper understanding of the course if they are to obtain the best results. They will need to do more than the minimum expected and some extra reading will be an important step in developing their appreciation of chemistry.

Students will have to prepare for questions that expect a deeper understanding. For this reason, each chapter ends with an extended question that is more demanding than the other problems. It is recognised that these questions are not relevant to all students and is anticipated that only a few will feel confident enough to tackle them without guidance. Some teachers might find that the best approach is to use these questions as a basis for class discussion.

Margin comments are provided throughout the book. These comprise valuable reminders and snippets of information and include examiner's tips (indicated by an **e** symbol).

# Required previous knowledge

Little previous knowledge is assumed, although the OCR specification is built on the understanding that students will have achieved a minimum of a grade C in either GCSE Science and Additional Science or GCSE Chemistry.

Some facility with basic mathematics is a prerequisite. In particular, students should be:
- comfortable with the use of a calculator
- able to manipulate simple algebraic equations to determine an unknown
- able to express numbers in scientific notation, for example to recognise that 0.00739 is equivalent to $7.39 \times 10^{-3}$.

It is hoped that the inclusion of a range of worked examples and simple mnemonics will help those who find the mathematical requirements of AS a burden.

CD-ROM
The enclosed CD-ROM has two components:
- end-of-chapter worksheets for students to print off and complete in order to create sets of summary notes
- exam-style questions with graded student answers, supported by examiner's comments

# Scheme of assessment

Candidates for AS take two written papers and a practical component that is internally assessed.

**Unit 1: Atoms, bonds and groups** is examined by a written paper, lasting 1 hour. The paper is worth 60 marks and carries 30% of the total marks for AS. All the questions are compulsory.

**Unit 2: Chains, energy and resources** is examined by a written paper, lasting 1 hour 45 minutes. The paper is worth 100 marks and carries 50% of the total marks for AS. All the questions are compulsory.

The papers cover the range of topics within those units and are constructed to test a candidate's knowledge and understanding as well as the ability to apply knowledge in an unfamiliar situation.

The internally assessed practical component requires more than just the ability to perform an experiment. Candidates have to provide evidence of:
- accuracy in observation and recording
- ability to interpret results, both qualitatively and quantitatively
- judgement in evaluating an experiment and providing possible improvements to the method employed

The practical component carries 20% of the marks at AS.

# Advice to candidates

The answers for each examination question on the written papers at AS are written in the question booklet. Therefore, a particular number of lines is given for your anticipated response. The first thing to do is to check the number of marks available for an answer. This is given in brackets at the end of the question or part-question. This is more important than the number of available lines, because it indicates how many points you should make in your response. Some candidates express themselves more concisely in their answers (or just have smaller writing), so the size of the space is only a rough guide as to how much you need to write. If you exceed the space provided, double-check that you have addressed the question correctly. Remember that there is always the risk that if you write too much, you could contradict yourself. If an answer is worth 1 mark and you make more than one statement, you would gain the mark if both statements are correct, but not if one statement is correct and the other incorrect.

You may want to cross out an answer that you believe is incorrect, but it is unwise to do so unless you are sure you are replacing it with something more accurate. It is surprising how often examiners see something worthy of a mark within a crossed-out passage. You might feel that if what you have written is wrong, then the examiner will take a worse view of you. This is not the case. An examiner sticks to the marking scheme and makes no overall judgement about you.

Finally, heed the obvious advice to read the question carefully. It is distressing for an examiner to read an answer that would have scored many marks if it had been a response to a question other than the one asked. Remember that if a question has a long introduction, it usually contains information that either is essential in order to answer the question or is designed to make it easier. It is not unusual for examiners to read incorrect formulae in an answer, even though the formulae concerned are given in the question.

## Terms used in examination questions

You will be asked precise questions in the examinations, so you can save valuable time, as well as ensuring that you score as many marks as possible, by knowing what is expected. Terms used most commonly are explained below.

- **Define** — a formal statement or equivalent paraphrase is required. Precise definitions are given in this book. If you learn these, you should score all the marks available for definitions.
- **Explain what is meant by** — this normally implies that a definition should be given, together with some relevant comment on the significance or context of the term(s) concerned, particularly if two or more terms are included in the question. The amount of supplementary comment intended should be interpreted according to the number of marks available.
- **State** — give a concise answer with little or no supporting argument or explanation.
- **Describe** — state in words (using diagrams and/or equations as appropriate) the main points of the topic. It is often used with reference to either particular phenomena or particular experiments. The amount of description required depends on the number of marks available.
- **Name** — give the full systematic name, not the formula.
- **Identify** — give the name, formula or structure.
- **Write the formula** — a molecular formula will suffice.
- **Write the structural formula** — show clearly the position of functional groups, and be unambiguous. A molecular formula will *not* suffice,
- **Deduce/predict** — you are not expected to recall the answer, rather to make a logical connection between pieces of information/data. Such information may be given wholly in the question or may depend on answers extracted in an earlier part of the question. 'Predict' also implies a concise answer, with no supporting statement required.
- **Outline** — restrict the answer to essential detail only.
- **Suggest** — this may imply either that there is no unique answer or that you are expected to apply your general knowledge to a 'novel' situation that is an extension of the specification content.
- **Calculate** — a numerical answer is required. Working should be shown. Your answer should be set out in such a way that if a mistake is made, this can be tracked by the examiner and marks can be awarded for any correct consequential steps.
- **Sketch** — applied to diagrams, this means that a simple, freehand drawing is acceptable. Nevertheless, care should be taken over proportions and the important details should be labelled clearly.

# Unit **1**

# Atoms, bonds and groups

# Atoms, isotopes and electrons

## Chapter 1

In this chapter you will learn about some of the ideas that led to our current understanding of the structure of the atom. Atoms contain protons and neutrons in their nuclei; you will learn about some of their features and their role in providing most of the mass of an atom. However, the focus is on the third type of atomic particle — the electron. The nature and position of the electrons in atoms allow us to understand the formulae of compounds and are crucial to our appreciation of the way chemical reactions occur.

## Atoms

The idea that all matter consists of small particles was a widely held belief that was first stated in a precise form by John Dalton in the 1820s. Even though it was not an original idea, his hypothesis contained more detail than anything suggested previously.

Key points of Dalton's hypothesis were that:
- chemical elements are made up of particles called atoms, which can be neither created nor destroyed
- atoms are indivisible
- the difference between the atoms of any one element and another is that they have different masses
- a compound is made up of atoms combined in a fixed ratio

Today we understand atomic structure more fully. Look at each of the key points above and decide whether it is still valid or needs modifying.

A brief historical sequence of how we moved from Dalton's theory to our present understanding is given on pages 19–21 under the heading *Additional reading*.

### Subatomic particles

A more detailed description of the atom was proposed early in the twentieth century following work carried out initially by Rutherford and his research team. Although it is now known that the structure of the atom is more complicated, you need to learn the following details:
- An atom consists of a nucleus that is made up of positively charged particles called **protons** and uncharged particles called **neutrons**.
- Negatively charged particles called **electrons** orbit the nucleus.

**e** Knowledge of this historical sequence is not required for the OCR specification and you will not be tested on it in an examination. However, it may improve your understanding and help you to answer questions more fully.

- Protons and neutrons have identical masses. Electrons are much lighter, being approximately 1/2000 of the mass of a proton or neutron.
- All atoms are neutral. Therefore, the number of electrons is equal to the number of protons.

It is worth appreciating that the diameter of the nucleus is only about 1/100 000 of the total diameter of the atom. So, if the nucleus were the size of an average bedroom in a house in Birmingham (Figure 1.1), the electron would follow roughly an orbit through Blackpool, Swansea, London and Skegness. An atom has a large amount of space between its nucleus and the electrons.

**Figure 1.1**
*A geographical illustration of the scale of an atom*

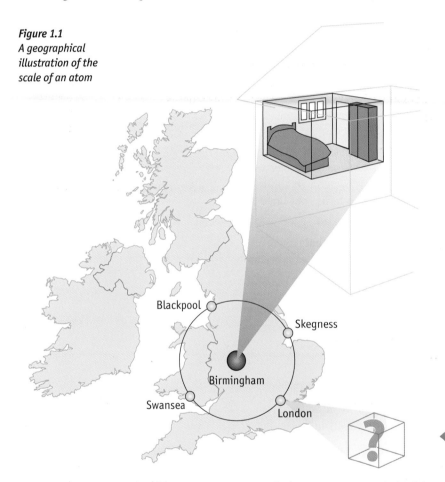

No one knows the size of an individual electron. As explained later, it cannot even be fully described as a particle.

The charges and masses of protons, neutrons and electrons are summarised in Table 1.1.

| Particle | Charge/coulombs | Mass/g |
|---|---|---|
| Proton (p) | $1.60 \times 10^{-19}$ | $1.67 \times 10^{-24}$ |
| Neutron (n) | 0 | $1.67 \times 10^{-24}$ |
| Electron (e) | $-1.60 \times 10^{-19}$ | $9.11 \times 10^{-28}$ |

**Table 1.1**
*Characteristics of subatomic particles*

You are not expected to recall the data given in Table 1.1. It is, however, important that you know the relative charges and masses of the particles (Table 1.2).

| Particle | Relative charge | Relative mass |
|---|---|---|
| Proton (p) | +1 | 1 |
| Neutron (n) | 0 | 1 |
| Electron (e) | −1 | 1/1833 ($\approx$1/2000) |

*Table 1.2 Relative charges and masses of subatomic particles*

- The atoms of any one element are different from the atoms of all other elements.
- All atoms of the same element contain the same number of protons; atoms of different elements contain a different number of protons.
- The number of protons contained in the atom of an element is its **atomic number**.
- The sum of the numbers of protons and neutrons in an atom is its **mass number**. This implies that the protons and neutrons supply the total mass of an atom. This is not completely true, because there are also electrons present. However, because electrons are so much lighter than protons and neutrons, it is a reasonable approximation.

It is easy to work out the numbers of particles present in an atom using its atomic number and mass number.

◁ The atomic number is sometimes referred to as the **proton number.** The mass number is sometimes referred to as the **nucleon number.**

### Worked example
An atom has atomic number 13 and mass number 27. How many of each subatomic particle does it contain?

### Answer
The atomic number is the number of protons, which is, therefore, 13.
Since an atom is neutral, the number of electrons is also 13.
The number of protons plus the number of neutrons is 27. Therefore, the number of neutrons is 14.

(The element is aluminium and it is usually written as $^{27}_{13}$Al.)

◁ Mass number — $^{27}_{13}$Al
Atomic number

Subtract the bottom number from the top number to get the number of neutrons.

## Formation of ions

Atoms form **ions** by losing or gaining electrons. In fact, losing or gaining electrons is the process by which many chemical reactions take place.

When an atom loses one electron, the number of electrons in the atom becomes one fewer than the number of protons. Therefore, it has an electrical charge of +1. For example, a sodium atom loses an electron, forming an ion that is represented by $Na^+$. A calcium atom can lose two electrons and, as it then has two fewer electrons than it does protons, it is written as $Ca^{2+}$.

If one electron is gained, the atom of the element contains one more electron than it does protons and, therefore, has an electrical charge of −1. For example, a chlorine atom gains an electron, forming a chloride ion, which is represented by Cl⁻. An oxygen atom gains two electrons, forming an oxide ion. This has two more electrons than protons and is written as $O^{2-}$.

Ions have different properties from those of the element from which they came. For example, sodium metal is extremely reactive but sodium ions are not.

◀ For non-metals, the element and its ion do not have the same name. A non-metallic ion has the ending '-ide'. For example, the ion from oxygen is oxide, from bromine it is bromide and from sulphur it is sulphide.

---

### Worked example 1

How many of each subatomic particle are present in a sulphide ion, $S^{2-}$? Sulphur has atomic number 16 and mass number 32.

**Answer**

Sulphur has atomic number 16, so a sulphur atom contains 16 protons.
Since the electrical charge of a sulphide ion is '2−', it contains two more electrons than it does protons. Therefore, the number of electrons is $16 + 2 = 18$.
The atomic mass is 32. The number of neutrons is the mass number minus the number of protons $(32 − 16)$, which is 16.
Therefore, a sulphide ion contains 16 protons, 16 neutrons and 18 electrons. It is written as $^{32}_{16}S^{2-}$.

---

### Worked example 2

How many of each subatomic particle are present in a potassium ion, $K^+$? Potassium has atomic number 19 and mass number 39.

**Answer**

Potassium has atomic number 19, so a potassium atom contains 19 protons.
Since the electrical charge of a potassium ion is '1+', it contains one more proton than it does electrons. Therefore, the number of electrons is 18.
The atomic mass is 39. The number of neutrons is $39 − 19$, which is 20.
Therefore, a potassium ion contains 19 protons, 20 neutrons and 18 electrons. It is written as $^{39}_{19}K^+$.

---

## Questions

Use the atomic numbers and mass numbers given to deduce the number of protons, neutrons and electrons present in the following atoms or ions:

**a** (i) Oxygen, O (atomic number 8, mass number 16)
  (ii) $O^{2-}$

**b** (i) Sodium, Na (atomic number 11, mass number 23)
  (ii) $Na^+$

**c** (i) Fluorine, F (atomic number 9, mass number 19)
  (ii) $F^-$

**d** (i) Calcium, Ca (atomic number 20, mass number 40)
  (ii) $Ca^{2+}$

# Isotopes and mass spectra

Isotopes are atoms of the same element that have different masses. The isotopes of an element have the same number of protons (and electrons), but different numbers of neutrons.

## Evidence for the existence of isotopes

Evidence for the existence of isotopes was obtained by using a mass spectrometer. In the 1920s, Francis Aston invented a machine that used the ionisation of elements and compounds to determine their individual masses. A gaseous sample of the substance is bombarded with high-energy electrons, which creates positive ions. These are separated according to their charge and those with a 1+ charge are selected. After being accelerated by an electric field, a focused stream of these ions is then passed at constant speed through a magnetic field. Here, the ions are deflected. The degree of the deflection depends on the mass of the ion. A detector is calibrated to record the degree of deflection and interpret this in terms of mass. This simple description does not do justice to the complex nature of the instrument. However, it is sufficient for you to understand the general principles without the precise details. The key features are summarised in Figure 1.2. In order to avoid interference from gases in the air, the spectrometer is evacuated using a pump.

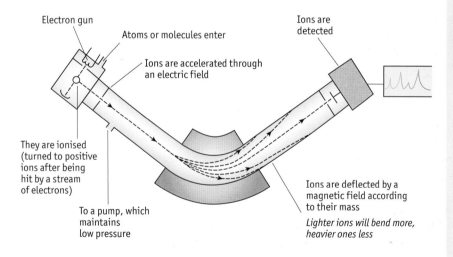

Electron gun

Atoms or molecules enter

Ions are accelerated through an electric field

Ions are detected

They are ionised (turned to positive ions after being hit by a stream of electrons)

To a pump, which maintains low pressure

Ions are deflected by a magnetic field according to their mass

*Lighter ions will bend more, heavier ones less*

*Figure 1.2 A representation of a mass spectrometer*

ⓔ Questions on mass spectrometry are set only on the second unit at AS. These details are provided to aid your understanding of the structure of atoms.

It is important to remember that the mass spectrometer analyses positive ions only. Do not confuse these ions with the ions that are formed under normal laboratory conditions. The conditions employed in the apparatus are such that it is impossible to create negative ions.

## Mass spectra

For an atom, a typical print-out from a mass spectrometer looks like a 'stick-diagram'. Each stick represents an ion (isotope); the taller the 'stick', the more abundant the ion.

ⓔ You will not be asked to describe a mass spectrometer in an exam. However, you need to understand how the results obtained from a mass spectrometer are used.

The mass spectrum for boron is shown in Figure 1.3.

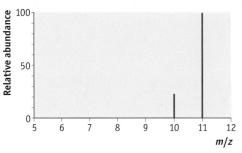

*Figure 1.3 The mass spectrum of boron*

The x-axis is labelled 'm/z', which means mass/charge. However, since the charge is 1+, it is effectively the relative mass of the ions that is recorded. A mass spectrometer is calibrated so that the isotope carbon-12 is given a value of exactly 12. This is the standard used universally for the comparison of the masses of atoms (see page 45). The y-axis records relative abundance, which is the number of times an ion of that mass was detected.

*A mass spectrometer*

The mass spectrum of boron has two lines. They are both a result of the formation of B$^+$ ions. One line is equivalent to a mass of 10 and the other to a mass of 11. This occurs because boron consists of two isotopes, the atoms of one isotope containing five neutrons and the other containing six neutrons. The relative abundance of these two isotopes is shown by the heights of the lines — they are in the ratio of 23 (boron-10) to 100 (boron-11). Knowing this ratio allows the mean relative atomic mass to be determined.

The word 'relative' means that the figures are compared with the mass of carbon-12.

Consider a total of 123 boron particles — 23 will be boron-10 and 100 will be boron-11. Therefore, the atomic mass has a contribution of (23/123) from boron-10, which is (23/123) × 10 = 1.87, and a contribution of (100/123) from boron-11, which is (100/123) × 11 = 8.94.

The mean relative atomic mass of boron is 1.87 + 8.94 = 10.81.

Strictly, it is better to use the expression *weighted* mean relative atomic mass. The ordinary mean of 10 and 11 is 10.5. However, here the numbers of atoms of each isotope are considered and, in these circumstances, the word 'weighted' is included.

In some problems the percentage abundances obtained from a mass spectrum are supplied. The following example shows how the atomic mass is then obtained.

### Worked example
Silver has two isotopes: 51.35% of the atoms are silver-107 and 48.65% of the atoms are silver-109. Calculate the weighted mean relative atomic mass of silver.

### Answer
The fraction that is $^{107}Ag^+$ is 51.35/100; the fraction that is $^{109}Ag^+$ is 48.65/100.

$$\text{weighted mean relative atomic mass} = \left(\frac{51.35}{100} \times 107\right) + \left(\frac{48.65}{100} \times 109\right) = 107.97$$

### Developments in mass spectrometry
The mass spectrometer is also used to analyse compounds, although the number of lines obtained may be large. This is because bombardment by electrons causes the molecule to break up and each of the fragments obtained registers on the detector. This can be an advantage, as it is sometimes possible to obtain details of the molecule's structure as well as its overall molecular mass. This need not be considered further here.

Whereas the original machine designed by Aston was large and only moderately accurate, today's instruments are much smaller and the results are accurate to four decimal places. This precision means that, for example, a mass spectrometer can distinguish between carbon monoxide, nitrogen and ethene even though to a first approximation each has a mass of 28. The accurate figures are: carbon monoxide, 28.0101; nitrogen, 28.0134; ethene, 28.0530.

A mass spectrometer uses very small samples of gaseous molecules and, today, can be operated remotely. This makes it a useful piece of equipment for analysing samples in remote locations. For example, the equipment on board the Mars space probe included a mass spectrometer that analysed samples of gas and returned the results to Earth.

**1** Rubidium consists of two isotopes. Rb-85 has an abundance of 72.2%; Rb-87 has an abundance of 27.8%.

Calculate the weighted mean relative atomic mass of rubidium.

**2** Neon consists of three isotopes. The percentage composition is Ne-20, 90.48%; Ne-21, 0.27%; Ne-22, 9.25%.

Calculate the weighted mean relative atomic mass of neon.

**3** Copper has a weighted mean relative atomic mass of 63.6. It contains two isotopes. One isotope has a mass of 63 and an abundance of 69%. Deduce the mass of the other isotope.

# Electrons

You may be familiar with a picture of atoms based on a model suggested by the Danish scientist, Niels Bohr. Each atom has electrons in orbits that are outside the nucleus and are of increasing size. The first orbit can accommodate two electrons, the second eight electrons and so on. Such diagrams show what is known as the **ground state** of the element. They are useful as a basic representation of atomic structure and help us to understand the nature of chemical bonding (Chapter 6). However, their use is limited to atoms with relatively few electrons.

## Ground states of some elements

The simplest atom is hydrogen. With atomic number 1, an atom of hydrogen contains one proton and hence one electron.

The next atom, helium, has atomic number 2. Therefore, there are two protons in its nucleus. The first orbit can accommodate two electrons only and it is therefore completely filled.

An atom of lithium has three protons in its nucleus. The third electron is in a new, larger orbit.

The second orbit can accommodate eight electrons. As the atomic number increases, the electrons progressively fill this orbit. This is illustrated by beryllium and carbon. Beryllium, $_4$Be, has four electrons arranged 2, 2; carbon, $_6$C, has six electrons arranged 2, 4.

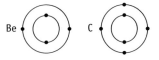

Neon, element 10, contains the maximum eight electrons in the second orbit. The next element, sodium, $_{11}$Na, utilises a third orbit. Its electrons are arranged 2, 8, 1.

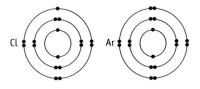

Chlorine and argon are further examples of elements with electrons in the third orbit.

**e** The orbits are also referred to as electron shells.

## Questions

Draw the electrons in the orbits for each of the following elements:

**a** Oxygen (atomic number 8)

**b** Magnesium (atomic number 12)

**c** Silicon (atomic number 14)

## Ionisation energies

At GCSE level, you probably had to accept the idea of electron orbits (or shells as they are sometimes called) without any experimental evidence. Some evidence for the existence of these orbits can be obtained by plotting the energy required to remove electrons, one by one, from an atom. These energies are called **ionisation energies.** When successive ionisation energies of an element are plotted, the graph has the shape shown in Figure 1.4.

Chlorine has 17 electrons, arranged 2, 8, 7. There is a steady increase in the energy required to remove the electrons. This is to be expected because, as each electron is removed, the nucleus exerts an increasing pull on the remaining electrons. There is a noticeable jump in the energy required after the seventh and the fifteenth electrons have been removed. This is because at these two stages the electron is being removed from an orbit (shell) nearer to the nucleus.

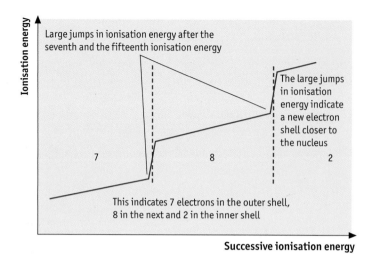

*Figure 1.4 Successive ionisation energies of chlorine, $_{17}Cl$*

In the figure: Large jumps in ionisation energy after the seventh and the fifteenth ionisation energy

The large jumps in ionisation energy indicate a new electron shell closer to the nucleus

This indicates 7 electrons in the outer shell, 8 in the next and 2 in the inner shell

**Successive ionisation energy**

### Evidence for the existence of subshells

Knowing the basic structure of the atom is helpful in understanding the way that atoms behave and you will use these ideas as your knowledge of chemistry develops. However, there are some aspects of the behaviour of electrons that cannot be explained by this model of the atom. Further research during the twentieth century revealed that electrons are more complicated than just small particles revolving around a central nucleus. First, there is experimental evidence to suggest that each shell is made up of smaller subshells.

By studying the first ionisation energy (the energy required to remove the first electron from an atom), it is found that the first 20 elements show a more complex pattern than might be expected (Figure 1.5). Trends in the ionisation energies of the elements are looked at in detail in Chapter 7. For now, notice that, although there is a general increase in ionisation energy until each shell is filled, the trend is not quite uniform. There are several small peaks and troughs. These peaks and troughs are repeated in each shell and provide some evidence that the arrangement of electrons involves subshells.

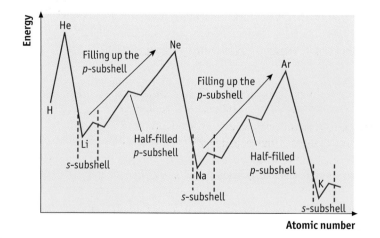

In the figure: He, Ne, Ar, H, Li, Na, K; Filling up the *p*-subshell; Half-filled *p*-subshell; *s*-subshell; **Atomic number**

*Figure 1.5 Evidence for the existence of subshells*

# Electron energy levels and orbitals

## Energy levels

Apart from the evidence obtained from the patterns of ionisation energies, other research led to the conclusion that instead of focusing on the position of electrons in terms of orbits it is more accurate to assign them energy levels (see *Additional reading* on page 20). These energy levels are numbered 1, 2, 3 etc. and correspond broadly to the shells. All but level 1 are subdivided and given letters that characterise them in more detail. The letters used are *s*, *p*, *d* and *f*. The first two are indicated in Figure 1.5.

The lowest electron energy level in an atom is '1*s*'. This is the only option for level 1. Stepping upwards in energy, the next energy level is '2*s*'. However, level 2 has a further energy subdivision called '2*p*'. Level 3 has 3*s*, 3*p* and also 3*d*. Level 4 has 4*s*, 4*p*, 4*d* and 4*f*.

An energy level labelled '*p*' is of higher energy than the equivalent '*s*' level. Therefore, 2*p* has higher energy than 2*s*; 3*p* has higher energy than 3*s* and so on.

It is important to know the order of increasing energy of the electron levels because this indicates the electron arrangement of the element in its ground state (i.e. the lowest energy state). These energies do not always follow the sequence you might expect, but fortunately there is an easy way of remembering the order (Figure 1.6).

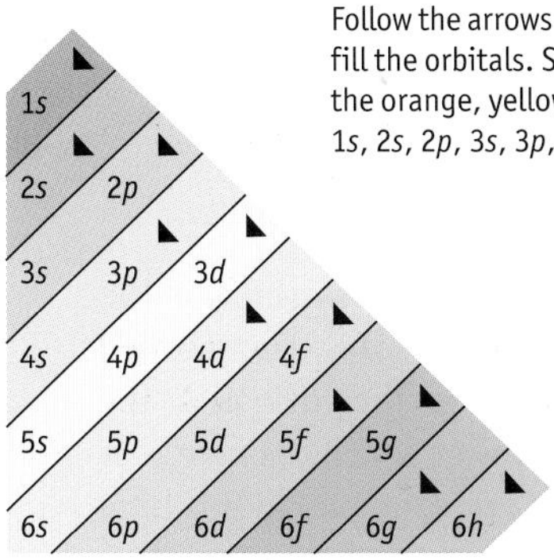

The numbers 1, 2, 3 etc. are called **principal quantum numbers**.

It may seem strange to use the letters *s*, *p*, *d* and *f*, but they relate to the names of the lines in the emission spectra of alkali metals:

- *s* = sharp
- *p* = principal
- *d* = diffuse
- *f* = fundamental

Emission spectra measure the small bursts of energy resulting from electrons moving within the atom.

Follow the arrows to obtain the order that the electrons fill the orbitals. Start with the red and follow with the orange, yellow, green etc. to give the sequence, 1*s*, 2*s*, 2*p*, 3*s*, 3*p*, 4*s*, 3*d*, 4*p*, 5*s*, 4*d*, 5*p* etc.

*Figure 1.6 Key to remembering the order in which orbitals are filled*

Note that the 4*s* fills up before the 3*d*; likewise, the 5*s* fills up before the 4*d*.

## Electron orbitals

The energy of an electron is no longer associated with a particular distance from the nucleus. You might, therefore, expect that it would be impossible to provide a visual representation of electrons in the various energy levels. This turns out not to be the case, although it does require a degree of imagination to understand it. The idea of the electron as a fixed particle in an orbit has to be replaced by the concept of it as a distribution of negative charge. If you had a single hydrogen atom and mapped the position of its one electron at regular intervals you could build up a three-dimensional map of places where the electron is likely to be found.

In the case of hydrogen, the electron in a 1s-orbital can be found anywhere within a sphere surrounding the nucleus. Figure 1.7 shows a cross section through this spherical space. For most of the time the electron is within a fairly easily defined region of space close to the nucleus. Such a region of space is called an **orbital**.

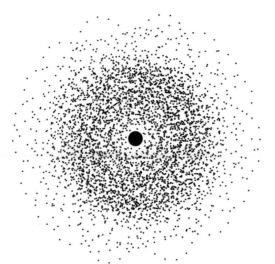

**Figure 1.7**
*A 1s-orbital*

### s-orbitals

Figure 1.7 shows an **s-orbital** and all s-orbitals have similar patterns. They consist of a spherical distribution of negative charge centred on the nucleus and radiating outwards with variations in density.

The most significant difference between a 1s-orbital and a 2s-orbital is the distance from the nucleus to where the major density of charge is concentrated (Figure 1.8). In the 1s-orbital, the electron density is closer to the nucleus. As the levels (principal quantum numbers) increase, the radius at which the charge is most dense becomes further away from the nucleus. For example, a 4s-orbital is larger than a 3s because the point at which the 4s-orbital has its greatest density of negative charge is further from the nucleus than it is for the 3s-orbital.

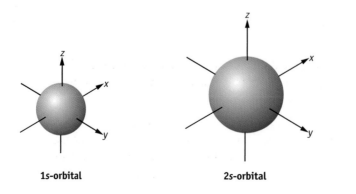

1s-orbital          2s-orbital

**Figure 1.8**
*The shape of
1s- and 2s-orbitals*

## p-orbitals

The **p-orbitals** are more complex and harder to visualise (Figure 1.9). They have an elongated dumbbell shape and a variable charge density, with the area of greatest concentration increasing with the distance from the nucleus as the principal quantum number increases. For each principal quantum number there are three p-orbitals. They are identical and have the same energy, differing only in their orientation in space. They are labelled '$x$', '$y$' and '$z$' to correspond to the three principal axes.

Orbitals that have the same energy are said to be **degenerate**.

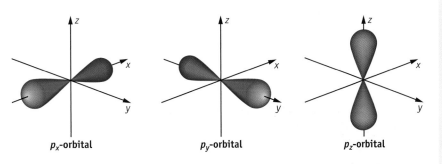

$p_x$-orbital          $p_y$-orbital          $p_z$-orbital

*Figure 1.9*
*The shape of 2p-orbitals*

## d-orbitals

The five **d-orbitals** are even more unusual. Their shapes are shown in Figure 1.10.

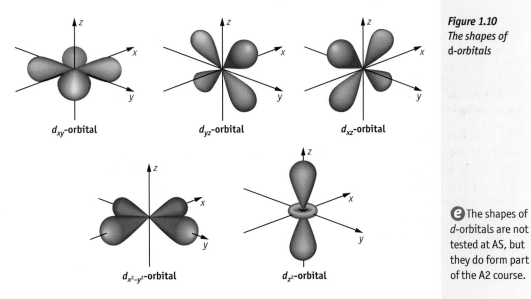

$d_{xy}$-orbital          $d_{yz}$-orbital          $d_{xz}$-orbital

$d_{x^2-y^2}$-orbital          $d_{z^2}$-orbital

*Figure 1.10*
*The shapes of d-orbitals*

e The shapes of d-orbitals are not tested at AS, but they do form part of the A2 course.

Although one of these d-orbitals looks noticeably different from the other four, they all have the same energy. The subscript labels come from the mathematical equations that define the orbitals.

Strictly, these representations are valid for the hydrogen atom only — it is assumed that the same pattern occurs with other elements. The validity of this assumption is reinforced by the way it helps us to understand further properties of the elements that you will encounter during the A-level course.

An orbital, whether *s*, *p*, *d* or *f*, is the region of space where there is the highest probability of finding an electron and can contain a maximum of two electrons. The orbitals are filled systematically, starting with the lowest energy level and building upwards.

## Electron configuration of the elements

Electron energy levels can be represented in 'box' diagrams, as shown in Figure 1.11.

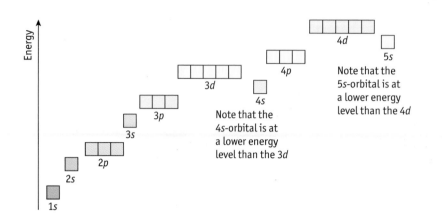

**Figure 1.11**
*Energy-level diagram*

Each box represents an orbital. Each orbital can contain a maximum of two electrons that have opposite spin. The orbitals are filled using an upward arrow (↑) to represent an electron spinning in one direction and a downward arrow (↓) for an electron spinning in the opposite direction.

Some examples of electron configurations using box diagrams are given in Figure 1.12.

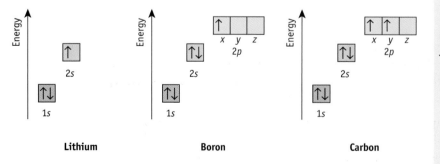

**Figure 1.12**
*Box diagrams for the electron configurations of lithium, boron and carbon*

The electron configurations of elements in the ground state can be derived using Figure 1.11.

- An atom of hydrogen has one electron, which in its ground state is in the 1*s*-orbital: $1s^1$.

- An atom of helium has two electrons that occupy the $1s$-orbital. Its electron configuration is written as $1s^2$.
- An atom of lithium has three electrons and its electron structure is written as $1s^2\,2s^1$.
- The electron structure of beryllium is $1s^2\,2s^2$.
- An atom of boron has five electrons and the next highest energy orbital, $2p$, is used. The electron configuration is written as $1s^2\,2s^2\,2p_x{}^1$, although the subscript $x$ is not necessary because any of the three $2p$-orbitals could be used as they all have the same energy. The label $x$ has no direct meaning until the $y$- and $z$-orbitals are also occupied. Labelling is only necessary to distinguish between orbitals of equivalent energy.

Each orbital of equivalent energy is occupied by one electron before the second electron is added. The reason for this is that two electrons within the same orbital experience a degree of repulsion that makes the pairing of electrons slightly less favourable.

- Carbon, atomic number 6, has the ground state $1s^2\,2s^2\,2p_x{}^1\,2p_y{}^1$. This is sometimes written as $1s^2\,2s^2\,2p^2$, i.e. the two $p$-orbitals are grouped together.
- Nitrogen, atomic number 7, has the ground state $1s^2\,2s^2\,2p_x{}^1\,2p_y{}^1\,2p_z{}^1$ or $1s^2\,2s^2\,2p^3$.

After nitrogen, the electrons in the $p$-orbitals have to pair up:
- Oxygen is $1s^2\,2s^2\,2p_x{}^2\,2p_y{}^1\,2p_z{}^1$ ($1s^2\,2s^2\,2p^4$).
- Neon is $1s^2\,2s^2\,2p_x{}^2\,2p_y{}^2\,2p_z{}^2$ ($1s^2\,2s^2\,2p^6$).

Once the $2p$-orbitals have been filled, the $3s$-orbital and the $3p$-orbitals are occupied in a similar way so that, for example, the ground state of silicon (14 electrons) is $1s^2\,2s^2\,2p^6\,3s^2\,3p^2$ and that of chlorine with 17 electrons is $1s^2\,2s^2\,2p^6\,3s^2\,3p^5$.

Note that once the $3p$-orbitals have been filled, the next orbital to be occupied is the $4s$ (*not* the $3d$). Therefore, the ground state of potassium (19 electrons) is $1s^2\,2s^2\,2p^6\,3s^2\,3p^6\,4s^1$.

The $3d$-orbitals are filled in the atoms of scandium (21 electrons) to zinc (30 electrons). These elements have a number of properties in common and are called **d-block elements**. It is usual to write the electron structure of the transition elements with the $4s$-orbital after the $3d$-orbital so that, for example, scandium would be written as $1s^2\,2s^2\,2p^6\,3s^2\,3p^6\,3d^1\,4s^2$.

## Questions

Give the electronic configuration for the following atoms or ions:

a Na          e $Ca^{2+}$

b K           f $Al^{3+}$

c N           g $Cl^-$

d $O^{2-}$    h $P^{3-}$

Box-diagram representations of the electronic ground states of some elements are given in Figure 1.13.

| Element | Number of electrons | 1s | 2s | 2p | 3s | 3p | 3d | 4s | 4p | 5s |
|---|---|---|---|---|---|---|---|---|---|---|
| Hydrogen | 1 | ↑ | | | | | | | | |
| Helium | 2 | ↑↓ | | | | | | | | |
| Lithium | 3 | ↑↓ | ↑ | | | | | | | |
| Carbon | 6 | ↑↓ | ↑↓ | ↑ ↑ | | | | | | |
| Neon | 10 | ↑↓ | ↑↓ | ↑↓ ↑↓ ↑↓ | | | | | | |
| Sodium | 11 | ↑↓ | ↑↓ | ↑↓ ↑↓ ↑↓ | ↑ | | | | | |
| Sulphur | 16 | ↑↓ | ↑↓ | ↑↓ ↑↓ ↑↓ | ↑↓ | ↑↓ ↑ ↑ | | | | |
| Argon | 18 | ↑↓ | ↑↓ | ↑↓ ↑↓ ↑↓ | ↑↓ | ↑↓ ↑↓ ↑↓ | | | | |
| Potassium | 19 | ↑↓ | ↑↓ | ↑↓ ↑↓ ↑↓ | ↑↓ | ↑↓ ↑↓ ↑↓ | | ↑ | | |
| Scandium | 21 | ↑↓ | ↑↓ | ↑↓ ↑↓ ↑↓ | ↑↓ | ↑↓ ↑↓ ↑↓ | ↑ | ↑↓ | | |
| Iron | 26 | ↑↓ | ↑↓ | ↑↓ ↑↓ ↑↓ | ↑↓ | ↑↓ ↑↓ ↑↓ | ↑↓ ↑ ↑ ↑ ↑ | ↑↓ | | |
| Zinc | 30 | ↑↓ | ↑↓ | ↑↓ ↑↓ ↑↓ | ↑↓ | ↑↓ ↑↓ ↑↓ | ↑↓ ↑↓ ↑↓ ↑↓ ↑↓ | ↑↓ | | |
| Bromine | 35 | ↑↓ | ↑↓ | ↑↓ ↑↓ ↑↓ | ↑↓ | ↑↓ ↑↓ ↑↓ | ↑↓ ↑↓ ↑↓ ↑↓ ↑↓ | ↑↓ | ↑↓ ↑↓ ↑ | |
| Krypton | 36 | ↑↓ | ↑↓ | ↑↓ ↑↓ ↑↓ | ↑↓ | ↑↓ ↑↓ ↑↓ | ↑↓ ↑↓ ↑↓ ↑↓ ↑↓ | ↑↓ | ↑↓ ↑↓ ↑↓ | |
| Strontium | 38 | ↑↓ | ↑↓ | ↑↓ ↑↓ ↑↓ | ↑↓ | ↑↓ ↑↓ ↑↓ | ↑↓ ↑↓ ↑↓ ↑↓ ↑↓ | ↑↓ | ↑↓ ↑↓ ↑↓ | ↑↓ |

**Figure 1.13**
*The electronic ground states of some elements*

# Summary

You should now be able to:

- describe protons, neutrons and electrons in terms of relative charge, relative mass and distribution
- define the terms atomic (proton) number, mass (nucleon) number and isotope
- deduce the numbers of protons, neutrons and electrons in atoms and ions
- calculate weighted mean mass from a mass spectrum
- write electronic configurations in terms of orbitals for the atoms and ions of the first 36 elements
- describe the shapes of *s*- and *p*-orbitals

# Additional reading

Dalton's ideas about atoms provided the foundation for work undertaken during the nineteenth century. However, by the start of the twentieth century, further experimentation began to throw doubt on some of his key ideas. Experimental results indicated that atoms were not the basic units of an element but could be broken into smaller particles. J. J. Thomson discovered that, when they were heated strongly, metals gave off tiny, negatively charged particles. A stream of these particles could be deflected by magnetic and electrical fields. He concluded that they resulted from the breakdown of the metal atoms. Further work established that the mass of one of these particles was approximately 1/2000 of the mass of the lightest atom, hydrogen. In addition, the discovery of radioactivity by the Curies suggested that some atoms were capable of discharging heavier, positively charged alpha particles.

Ernest Rutherford carried out an experiment in which alpha particles were used to bombard gold foil. It was found that, although a few bounced back, most of the alpha particles passed through the foil. This suggested that an atom, rather than being a uniform solid particle, has a small, hard central core surrounded by a large volume of matter so diffuse that some particles pass through it. Thus, a new vision of the atom was proposed by Rutherford, working with a team of scientists in Cambridge.

## Ernest Rutherford, 1871–1937

Rutherford was born near Nelson, New Zealand, the fourth of 12 children. He was academically successful at Nelson College, played in the first rugby 15 and eventually became head boy. He graduated from Canterbury College in Christchurch, briefly contemplated a career as a schoolteacher, but travelled instead to England. His brilliance was soon recognised; he was awarded the Nobel prize for chemistry in 1908. He was rather offended that it was not in physics, which he regarded as a superior discipline. He is said to have once remarked that 'all science is either physics or stamp collecting'. Apart from this misjudgement, he was undoubtedly one of the greatest scientists of all time. Einstein called him 'the second Newton'. He was knighted in 1914, appointed the Director of the Cavendish laboratory in Cambridge in 1919 and became Lord Rutherford of Nelson in 1931.

TOPFOTO

## Electrons and electron orbitals

Rutherford was not certain how electrons were able to orbit the nucleus. It seemed to defy logic that they could continue circling without plunging into the nucleus because the nucleus exerts a strong force of attraction. Neils Bohr used Rutherford's ideas to suggest a more detailed model of atomic structure.

It was soon recognised that, although Bohr's ideas contained an essential truth, the behaviour of electrons was more elaborate. They behaved in some ways like very small particles but also had properties characteristic of a beam of light. No one was sure of the best description.

The answer was, at least in part, provided by Erwin Schrödinger's analysis, based on a suggestion of Louis de Broglie. Light was known to have some unusual properties. Sometimes it behaved as though it were a stream of particles and sometimes as though it were a wave. For example, when light is reflected from the face of a mirror it behaves as a stream of particles would when bouncing back from a hard, flat surface. On other occasions, light appears to radiate outwards like the ripples formed when a stone is dropped into a pond. De Broglie proposed that this dual behaviour might also apply to electrons.

LIBRARY OF CONGRESS/SPL

## J. J. Thomson, 1856–1940

J. J. Thomson hoped to become an engineer and studied at what is now the University of Manchester. However, his parents could not afford the apprenticeship fees. He was awarded a scholarship and moved to Trinity College, Cambridge. In 1884, he was appointed Cavendish Professor of Experimental Physics. Reputedly, he was rather ham-fisted but he had a genius for experimental design and analysis. His discovery of the electron led to a Nobel prize in 1906. On his appointment as Master of Trinity College, he relinquished his professorship to his one-time pupil Ernest Rutherford. Thomson's honours included 21 honorary doctorates from universities around the world. He is buried close to Isaac Newton in Westminster Abbey.

### Niels Bohr, 1885–1962

Bohr was born into a wealthy family with a property overlooking Christianborg Castle, where the Danish parliament used to meet. At school he was a good, but not outstanding, student. He was an excellent

football player, although he did not achieve the fame of his brother, Harald, who played in the Danish team that won a silver medal in the 1908 Olympics. His brilliance shone through when he went to England on a Carlsberg travel grant. Initially, he joined J. J. Thomson's team in Cambridge but was soon in Manchester working with Rutherford's group. He returned to Copenhagen in 1916 as Professor of Physics. He established the Institute of Theoretical Physics in 1921 and was one of the stars of twentieth-century science. He won the Nobel prize for physics in 1922.

TOPFOTO

### Erwin Schrödinger, 1887–1961

Along with Max Born, Schrödinger provided the theoretical basis of our understanding of the nature of the electron. Although he once said of quantum physics, 'I don't like it and I am sorry I ever had anything to do with it', he continued all his life to contribute to the development of the theory. He disapproved of Nazi anti-Semitism and left Berlin in 1933 —

the same year that he was awarded a Nobel prize. He intended to settle in Oxford, but the university authorities were scandalised when they discovered that he lived with both his wife and a mistress. By 1940, he had made a home in Dublin, where he stayed for 17 years. He wrote many books, one of which — *What is Life?* — was said by James Watson to be the impetus that led him to discover the intricacies of DNA. Schrödinger had a life-long interest in Hindu philosophy.

MARY EVANS PICTURE LIBRARY/ALAMY

Schrödinger took this as a starting point and used it to provide a mathematical analysis of the electronic structure of the hydrogen atom. His complex calculations produced conclusions that are of great importance to our understanding of chemistry. He found that an electron is able to exist in a far wider range of energy states than had been thought previously and that it exists as a cloud of negative charge, rather than as a tiny particle. The electron cloud has a variable charge density and, sometimes, an unusual shape. The word 'state' rather than 'position' had to be used because his theory did not directly associate the energy of an electron with its distance from the nucleus of the atom.

There are many such states available to the electron. Since it was no longer possible to provide a simple diagram of the structure, the states were given labels derived from features of Schrödinger's mathematics.

Major energy states were recognised and indicated by the principal quantum numbers (1, 2, 3, 4 etc.). The energy increases as the number increases. Within these major groupings there are subdivisions indicated by letters (*s*, *p*, *d* and *f*). However, there are restrictions on which of these subdivisions are possible for a given principal quantum number.

Each of the energy states within a hydrogen atom represents one solution of the equation that Schrödinger proposed. It may seem puzzling that a mathematical equation can lead to more than one solution, but consider the case of $\sin x = 0$. This could mean that $x$ is 0°, 180°, 360°... There is not an exclusive solution. The answer could be expressed as $n(180°)$, where $n$ is any integer. This '$n$' is like the principal quantum number. Schrödinger's equation describes the three-dimensional picture of the electron in hydrogen. This makes it necessary to have three parameters to provide the multiple solutions to his equation. These correspond to other quantum numbers needed to define the energy state of the electron. Wolfgang Pauli added to Schrödinger's ideas by stating the important principle

### Louis de Broglie, 1892–1987

Louis de Broglie was born into an aristocratic French family and at first seemed destined for a diplomatic career. He graduated from the Sorbonne in 1910 with a degree in history. After much agonising, he switched

to theoretical physics with such effect that his contributions were recognised with a Nobel prize in 1929. He became professor of theoretical physics at both the Henri Poincaré Institute and at the Sorbonne. He published 25 scientific texts and wrote extensively on the philosophical nature of modern physics. On the death of his elder brother in 1960, he inherited the title the seventh Duc de Broglie.

MEGGERS GALLERY/AMERICAN INSTITUTE OF PHYSICS/SPL

that only one electron can exist in a particular energy state. If two electrons are present in the same orbital, this is only possible because they have opposing spins.

Note that a precise description of the energy levels can be obtained for the hydrogen atom only. For atoms that have more than one electron it has proved impossible to set up and solve the necessary equation. It is, therefore, an assumption supported by experimental evidence that a similar pattern will be observed for all other elements.

## Wolfgang Ernst Pauli, 1900–58

Pauli was born in Vienna in 1900 and was soon recognised as a prodigy. He was so successful with his studies that he was awarded a doctorate at the age of 21, having written a thesis on quantum theory related to ionised hydrogen. By 1928, he was Professor of Theoretical Physics in Zurich and making many profound contributions to the understanding of quantum mechanics, in particular the effect of electron spin. After a brief, unhappy marriage he had a nervous breakdown and corresponded with Carl Jung, with whom he exchanged many thoughts and ideas. At the outbreak of the Second World War, he moved to the USA and became a naturalised citizen. He was awarded a Nobel prize in 1945. Pauli's experimental skills were not particularly refined — his colleagues referred to a 'Pauli effect' that had nothing to do with electron spin but more to do with the breakage of equipment that seemed to occur whenever he was in the room.

## Extension question: f- and g-orbitals and einsteinium

You have already learned that s-, p- and d-orbitals can hold two, six and ten electrons respectively. You might have noticed that there is a pattern to the total number of electrons that can be held in the orbitals of each shell. For example, when $n = 1$, only the 1s energy level is possible and it holds two electrons. When $n = 2$, there are the 2s and 2p energy levels, so level 2 can hold a total of eight electrons. Level 3 has the 3d-orbitals, which means that level 3 can hold a total of 18 electrons. The total number of electrons that can be held at the various shells is given by the formula $2n^2$.

a  Deduce the total number of electrons that can be held in the fourth and fifth shells.

b  The extra numbers of electrons are possible because level 4 has f-orbitals and level 5 has f- and g-orbitals. How many different f- and g-orbitals are there?

In November 1952, the first large thermonuclear explosion was carried out in the Pacific Ocean. It was codenamed 'Mike' and was the prototype of the 'hydrogen bomb'. President Eisenhower of the USA commented that 'humanity has now achieved for the first time in its history, the power to end its history'.

In the fallout of that event, a new transuranic element was identified. The element is called einsteinium and has the symbol Es. It took until 1961 before enough of the element — 0.01 mg — had been made to study its properties. It was found that its atomic number was 99 and it had at least two isotopic forms, with mass numbers 253 and 254. Since then, it has been established that einsteinium has 14 isotopes.

c  Give the numbers of protons, neutrons and electrons found in an atom of the einsteinium-253 isotope.

d  What is the difference in the number of particles present in the atoms of einsteinium-253 and einsteinium-254?

e  Use your answer to part (a) to decide how many 5f-electrons there are in an atom of einsteinium.

All elements with such high atomic numbers have nuclei that break down spontaneously. This is the cause of radioactivity. Usually this breakdown (known as decay) occurs in several steps, ultimately producing the element lead. In the first step of its decay einsteinium releases alpha radiation, which consists of particles containing two protons and two neutrons.

f  Suggest what happens to an atom of einsteinium when it loses an alpha particle.

# Formulae and equations

## Chapter 2

Many experiments in chemistry require precise quantities of reagents. It is important to understand how these quantities are decided. This involves being able to write the correct formulae for simple compounds and using these formulae to produce balanced equations for chemical reactions. An equation does more than provide information about what reacts together and what is produced; it gives exact details of the numbers of particles — atoms, molecules or ions — that are involved.

This chapter should be studied carefully because much of what follows in this book depends on your ability to apply the ideas met here.

The last section, covering ionic equations (pages 30–34), can be studied now or left until the work on structure and bonding (Chapter 6) has been completed.

## Formulae

Electrons are found on the outside of atoms. This means that they are more readily removed from an atom than the protons and neutrons in the central nucleus. Chemical bonds are formed because atoms can lose, gain or share electrons with other atoms. The number of electrons that can be lost, gained or shared is usually limited, and which of these processes occurs depends on the type of atom. (See Chapter 6 for more detail.)

### Valency

Metal atoms usually lose electrons; non-metals usually gain or share electrons. The number of electrons involved is a characteristic property of the element known as its **valency**. For example, an atom of sodium can lose only one electron, so sodium has valency 1. An atom of oxygen can gain two electrons, so oxygen has valency 2.

A valency can also sometimes be given to groups of atoms. For example, a nitrate group ($NO_3$) consists of one nitrogen atom and three oxygen atoms combined together and this whole unit has valency 1.

Valency does not specify whether electrons are gained or lost, but it is a helpful concept that can be used to generate accurate formulae. The importance of being able to work out the formulae of compounds cannot be underestimated.

Table 2.1 is a list of the valencies that you should learn. The groups referred to in the table are those of the periodic table (page 264).

*A copy of the periodic table is provided in the examination.*

*Table 2.1 Valencies of elements and groups of elements*

| Valency | Element/group losing electrons | Element/group gaining electrons |
|---|---|---|
| 1 | All group 1 elements, hydrogen (H), silver (Ag), ammonium ($NH_4$) | All group 7 elements, hydroxide (OH), nitrate ($NO_3$), hydrogencarbonate ($HCO_3$) |
| 2 | All group 2 elements, iron (Fe), copper (Cu), zinc (Zn), lead (Pb), tin (Sn) | Oxygen (O), sulphur (S), sulphate ($SO_4$), carbonate ($CO_3$) |
| 3 | All group 3 elements, iron (Fe) | |
| 4 | Carbon (C), silicon (Si), lead (Pb), tin (Sn) | |

> You can use the groups of the periodic table to help you remember some of these valencies correctly.

### Using valency to write chemical formulae

The second column in Table 2.1 contains electron givers (mostly metals). Electron givers make compounds with the electron receivers (non-metals or groups of non-metals) listed in the third column. If each has the same valency, they combine directly. For example, the formula of potassium bromide is KBr (both potassium and bromine have valency 1) and the formula of barium oxide is BaO (both barium and oxygen have valency 2).

If the valencies differ, the numbers of the components have to be balanced until each of their valencies is satisfied. For example, sodium has valency 1 and oxygen has valency 2. In sodium oxide, the valency of oxygen must be satisfied by the use of two sodiums. The formula is $Na_2O$. The subscript 2 indicates that two sodiums are required for each oxygen. This balancing of valencies is essential. Sodium cannot form a compound of formula NaO. In terms of the electrons that are involved, each oxygen can react only by receiving two electrons and each sodium can provide only one electron. Therefore, it is necessary to have two sodiums for each oxygen.

> Elements with valency 1 are called **univalent**. Those with valency 2 are **divalent** and those with valency 3 are called **trivalent**.

- The formula of calcium chloride is $CaCl_2$, since two chlorines are required to balance the valency of calcium.
- The formula of hydrogen sulphide is $H_2S$, since the sulphur requires two hydrogens to balance its valency.
- Aluminium oxide is more difficult because aluminium has valency 3 and oxygen has valency 2. The solution is to double the aluminium and triple the oxygen, making the total valency of each element up to '6'. The formula is $Al_2O_3$. In terms of electrons, each of two aluminium atoms provides three electrons (total six) and each of three oxygen atoms receives two electrons.
- The formula of potassium sulphate is $K_2SO_4$, since the sulphate unit ($SO_4$) has valency 2 and potassium has valency 1.

Note that some units consist of several atoms linked together and form compounds with the unit kept intact. For example, sulphate consists of five atoms: one sulphur atom and four oxygen atoms joined together. This unit does not change when deriving the formula of a compound.

- The formula of lithium carbonate is $Li_2CO_3$, because the carbonate unit has valency 2 and lithium has valency 1.
- The formula of magnesium nitrate is $Mg(NO_3)_2$.

Notice the use of brackets in the formula for magnesium nitrate. Magnesium with valency 2 requires two nitrate units of valency 1. The whole nitrate unit must be kept intact and brackets are used to make this clear. They are essential to avoid the ambiguity of writing $MgNO_{32}$ — magnesium nitrate does not contain 32 oxygens and only one nitrogen.

- The formula of aluminium hydroxide is $Al(OH)_3$, since three hydroxide units of valency 1 are needed for each aluminium of valency 3. The brackets are essential because if the formula were written as $AlOH_3$, it would imply that there were three hydrogens but only one oxygen.

Acids are compounds containing hydrogen that has given one electron to an electron receiver. The formula of hydrochloric acid, which is a compound containing hydrogen and chlorine, is $HCl$ and that of sulphuric acid is $H_2SO_4$, as it is a combination of hydrogen (valency 1) and sulphate (valency 2).

Some elements can have more than one valency — for example, iron can have a valency of 2 or 3 (Table 2.1). The name given to the compounds reflects the valency of the element: for example, iron(II) chloride indicates a compound containing iron with valency 2; iron(III) sulphate contains iron with valency 3. The formulae are, therefore, $FeCl_2$ and $Fe_2(SO_4)_3$ respectively.

Another way of deducing formulae is to use the 'valency crossover' technique. Using aluminium oxide as an example, follow these simple steps.

**Step 1** Write each symbol:

Al O

**Step 2** Write the valency at the top right-hand side of each symbol:

$Al^3$ $O^2$

**Step 3** Cross over the valencies:

$Al^3_2 O^2_3$

**Step 4** Write the crossed-over valency at the bottom right of the other symbol:

$Al_2O_3$ is the formula

With some compounds, there is an additional step because the ratio of the component elements is simplified to give the lowest possible values, as shown below for silicon oxide.

**Step 1** Write each symbol:

Si O

**Step 2** Write the valency at the top right-hand side of each symbol:

$Si^4$ $O^2$

**Step 3** Cross over the valencies:

$Si^4_2 O^2_4$

**Step 4** Write the crossed-over valency at the bottom right of the other symbol:

$Si_2O_4$

**Step 5** Simplify to give the lowest possible ratio:

$SiO_2$ is the formula

## Ions

Compounds are formed either by sharing electrons between atoms or by transferring electrons from one atom to another (Chapter 6). In a compound formed from a metal and a non-metal, electrons are transferred from the metal to the non-metal. The valency indicates the number of electrons involved. This transfer of electrons results in the components of the compound becoming electrically charged. For example, when chlorine receives one electron, it has one more electron than it has protons. Consequently, it has an electric charge of −1. Suppose this electron has been provided by potassium. The potassium has one less electron and an electric charge of +1, which indicates that it has one more proton than it has electrons. Chlorine in the compound potassium chloride is therefore best written as $Cl^-$ and potassium as $K^+$. The particles are no longer atoms and are referred to as ions.

> An ion is an electrically charged particle formed by the loss or gain of one or more electrons from an atom or a group of atoms.

An oxygen atom needs to gain two electrons to form the ion $O^{2-}$. A sodium atom gives one electron, so its ion is $Na^+$. Two sodium atoms are required to provide the necessary electrons for each oxygen atom. Compounds of the type are called **ionic compounds**.

Compounds formed by the *sharing* of electrons do not contain ions — for example, water and carbon dioxide. Further examples involve non-metallic elements where the element exists naturally as a pair of atoms joined together, such as $H_2$, $O_2$, $N_2$, $Cl_2$, $Br_2$ and $I_2$.

◁ Atoms of the same element joined in pairs are **diatomic molecules**.

## Questions

1 Use valencies to explain the formulae of the following compounds:
   a  Lithium chloride, LiCl
   b  Magnesium oxide, MgO
   c  Potassium oxide, $K_2O$
   d  Barium iodide, $BaI_2$

2 Select the correct formula for the following compounds and explain your choice.
   a  Silver chloride: AgCl, $AgCl_2$
   b  Magnesium chloride: MgCl, $MgCl_2$, $MgCl_3$
   c  Calcium hydroxide: CaOH, $CaOH_2$, $Ca(OH)_2$
   d  Sodium sulphate: $NaSO_4$, $(Na)_2SO_4$, $Na_2SO_4$, $Na_2(SO)_4$
   e  Iron(III) chloride: $Fe_3Cl$, $FeCl_3$

3 Use the data in Table 2.1 to answer the following questions:
   a  The formula of strontium bromide is $SrBr_2$. Deduce the valency of strontium.
   b  The formula of sodium phosphide is $Na_3P$. Deduce the valency of phosphorus.
   c  Potassium chlorate contains the unit '$ClO_3$' and its formula is $KClO_3$. Deduce the valency of the chlorate unit.

ⓔ Remember that you need to learn all the valencies listed in Table 2.1.

**4** Write the correct chemical formula for each of the following compounds:
  **a** Magnesium chloride
  **b** Sodium hydroxide
  **c** Calcium carbonate
  **d** Nitric acid
  **e** Barium hydroxide
  **f** Silver sulphate
  **g** Sodium carbonate
  **h** Zinc nitrate
  **i** Nitrogen
  **j** Iron(II) bromide
  **k** Iron(III) oxide
  **l** Aluminium sulphate

**5 a** The formula of rubidium chloride is RbCl. What is the formula of rubidium sulphate?
  **b** The formula of manganese sulphate is $MnSO_4$. What is the formula of manganese bromide?
  **c** The formula of chromium chloride is $CrCl_3$. What is the formula of chromium nitrate?
  **d** The formula of silver phosphate is $Ag_3PO_4$. What is the formula of phosphoric acid?

**6** Write the formula for each of the following:
  **a** Lead(IV) oxide
  **b** Tin(II) chloride
  **c** Iron(III) sulphate

**7** Name the following compounds:
  **a** $Fe_2O_3$
  **b** $PbS$
  **c** $Sn(NO_3)_2$
  **d** $PbI_4$
  **e** $SnI_4$

**8** Write the formulae of the ions found in each of the compounds in question 4. For example, magnesium chloride contains the ions $Mg^{2+}$ and $Cl^-$.

# Equations

Being able to state the correct formulae for substances is important in writing equations for chemical reactions. An equation includes the reactants used and the products obtained and indicates the number of particles of each substance.

## Balancing equations

A simple example is the reaction of zinc and sulphur to make zinc sulphide. This is summarised in an equation as:

$$Zn + S \rightarrow ZnS$$

It means that one atom of zinc reacts with one atom of sulphur to make one particle of zinc sulphide.

Contrast this with the reaction between silver and sulphur. The formula of silver sulphide is $Ag_2S$, indicating there are two silver ions for each sulphide ion.

The equation is:

$$2Ag + S \rightarrow Ag_2S$$

The '2' before the symbol Ag indicates that two atoms of silver are required for each atom of sulphur in order to make one particle of silver sulphide. The process of putting the '2' in front of the silver is called **balancing the equation**. Notice that within formulae, the numbers of atoms required are shown by using a subscript (as in $Ag_2S$), but when balancing an equation the numbers are written before the formula of the substance (as in 2Ag).

Aluminium reacts with sulphur to give aluminium sulphide. Its formula is $Al_2S_3$ and the equation is:

$$2Al + 3S \rightarrow Al_2S_3$$

This indicates that two aluminium atoms and three sulphur atoms are required in order to make each particle of aluminium sulphide.

> *Don't let the word 'ion' put you off. Before they react, elements are made of atoms. Metals react by losing electrons and non-metals by gaining electrons, so the substance formed contains ions.*

---

### Worked example 1

Write a balanced equation for the reaction of magnesium and oxygen to make magnesium oxide.

**Answer**

First, write the correct formulae for the reactants and products:

$$Mg + O_2 \rightarrow MgO$$

Now examine the formulae to decide what numbers of each particle are required and how much product is formed. Oxygen is $O_2$ and the formula of magnesium oxide is MgO, so two particles of MgO are made. Therefore, the balanced equation is:

$$2Mg + O_2 \rightarrow 2MgO$$

When balancing an equation, do not alter the formulae of the reactants or products. These are fixed by the valency rules and must not be changed.

> *Remember that oxygen is $O_2$ (not O) and that magnesium oxide is MgO.*

---

### Worked example 2

Write a balanced equation for the reaction between hydrogen and oxygen to form water.

**Answer**

The unbalanced equation is:

$$H_2 + O_2 \rightarrow H_2O$$

The formulae of oxygen and water are $O_2$ and $H_2O$ respectively. Therefore, two particles of water are made and $2H_2$ are required. Therefore, the balanced equation is:

$$2H_2 + O_2 \rightarrow 2H_2O$$

> *It is also acceptable to use fractions to balance the equation:*
> $$H_2 + \tfrac{1}{2}O_2 \rightarrow H_2O$$

### Worked example 3

Write a balanced equation for the reaction between sodium hydroxide and hydrochloric acid to form sodium chloride and water.

**Answer**

Writing down the correct formulae of the reactants and products gives:

$$NaOH + HCl \rightarrow NaCl + H_2O$$

On inspection, this equation is already balanced correctly.

When balancing more complicated equations, check each atom in turn to confirm that the number used on the left-hand side of the equation is the same as the number obtained on the right-hand side. In worked example 3, note that the two hydrogen atoms required to form the water come from different sources — one from sodium hydroxide and one from hydrochloric acid.

### Worked example 4

Write a balanced equation for the reaction between sodium hydroxide and sulphuric acid to form sodium sulphate and water.

**Answer**

The unbalanced equation is:

$$NaOH + H_2SO_4 \rightarrow Na_2SO_4 + H_2O$$

Balancing is required, as follows:

$$2NaOH + H_2SO_4 \rightarrow Na_2SO_4 + 2H_2O$$

The formula of sodium sulphate indicates that 2NaOH are required.

### Worked example 5

Write a balanced equation for the reaction between aluminium and sulphuric acid to form aluminium sulphate and hydrogen.

**Answer**

The unbalanced equation is:

$$Al + H_2SO_4 \rightarrow Al_2(SO_4)_3 + H_2$$

Balancing this equation is trickier. Note that to form the aluminium sulphate, 2Al are required and $3H_2SO_4$ are needed to provide the sulphate. The balanced equation is:

$$2Al + 3H_2SO_4 \rightarrow Al_2(SO_4)_3 + 3H_2$$

### Worked example 6

Write a balanced equation for the reaction between potassium carbonate and hydrochloric acid to form potassium chloride, carbon dioxide and water.

**Answer**

The unbalanced equation is:

$$K_2CO_3 + HCl \rightarrow KCl + CO_2 + H_2O$$

It is easy to see that 2KCl must be formed, so 2HCl are needed. Count the oxygen atoms, which are divided between the carbon dioxide and the water, and it can be seen that the balanced equation is:

$$K_2CO_3 + 2HCl \rightarrow 2KCl + CO_2 + H_2O$$

In summary, the procedure to be adopted in balancing equations is:
- Use valencies to determine the correct formula of each reactant and product.
- Write down the unbalanced equation with the reactants on the left-hand side of the arrow and the products on the right-hand side.
- Make sure that the number of each atom on the left-hand side equals the number on the right-hand side.

## Questions

**1** Balance the following equations:
  **a** $Zn + O_2 \rightarrow ZnO$
  **b** $Na + Cl_2 \rightarrow NaCl$
  **c** $CaO + HCl \rightarrow CaCl_2 + H_2O$
  **d** $Zn + AgNO_3 \rightarrow Zn(NO_3)_2 + Ag$
  **e** $Fe + Cl_2 \rightarrow FeCl_3$

**2** For each of the following reactions, first write an unbalanced equation and then amend it to produce a balanced equation.
  **a** sodium + oxygen → sodium oxide
  **b** zinc + hydrochloric acid → zinc chloride + hydrogen
  **c** hydrogen + copper oxide → copper + water
  **d** potassium hydroxide + nitric acid → potassium nitrate + water
  **e** calcium hydroxide + hydrochloric acid → calcium chloride + water
  **f** sodium + water → sodium hydroxide + hydrogen
  **g** chlorine + potassium iodide → potassium chloride + iodine
  **h** magnesium carbonate + hydrochloric acid → magnesium chloride + carbon dioxide + water
  **i** silver nitrate + copper chloride → silver chloride + copper nitrate
  **j** iron(II) oxide + nitric acid → iron(II) nitrate + water
  **k** iron(III) chloride + sodium hydroxide → iron(III) hydroxide + sodium chloride
  **l** magnesium + aluminium sulphate → magnesium sulphate + aluminium

**3** The chemical industry processes large amounts of limestone (calcium carbonate). It is heated to produce calcium oxide for use in cement production and in neutralising excess acidity in soils (liming). Carbon dioxide is given off when limestone is heated.

Calcium oxide is also used in a reaction with carbon to form an unusual compound, calcium carbide, $CaC_2$, with carbon dioxide as a by-product. Calcium carbide, $CaC_2$, reacts with water to produce the flammable gas ethyne (acetylene), $C_2H_2$.

Calcium oxide reacts with water to form calcium hydroxide. Calcium hydroxide reacts with chlorine to produce an aqueous paste of bleaching powder. This is a mixture of water, calcium chloride and calcium hypochlorite, $Ca(OCl)_2$.

Calcium hydroxide is also used for liming and, in one industrial process, in a reaction with sodium carbonate to produce sodium hydroxide and a precipitate of calcium carbonate. Calcium hydroxide is also formed.

Write equations for the following reactions:
**a** The action of heat on calcium carbonate
**b** The reaction between water and calcium oxide
**c** The reaction between calcium hydroxide and sodium carbonate
**d** The reaction between calcium oxide and carbon
**e** The reaction between calcium carbide and water
**f** The production of bleaching powder

## State symbols

Equations can be made more informative by including state symbols. The equation for the reaction between silver nitrate and potassium bromide is as follows:

$$AgNO_3 + KBr \rightarrow AgBr + KNO_3$$

This is correctly balanced, but it does not indicate that the reaction occurs only if the silver nitrate and potassium bromide are both in aqueous solution. Nor does it give the useful information that mixing the solutions produces solid silver bromide. The physical state of the substances in the equation can be shown by including state symbols:

$$AgNO_3(aq) + KBr(aq) \rightarrow AgBr(s) + KNO_3(aq)$$

The symbol (aq) indicates aqueous solution; (s) indicates that the silver bromide is a solid.

◀ 'Aqueous solution' means dissolved in water.

There are four state symbols:
- (s) for a solid
- (l) for a liquid
- (g) for a gas
- (aq) for an aqueous solution

The equation for the reaction of calcium carbonate with dilute hydrochloric acid uses all four state symbols:

$$CaCO_3(s) + 2HCl(aq) \rightarrow CaCl_2(aq) + CO_2(g) + H_2O(l)$$

# Ionic equations

This section may be more fully understood once Chapter 6 on bonding and structure has been studied. You may decide to return to it once you have completed Chapter 6.

Ionic equations are used to draw attention to the particles that react and ignore those that, although present, take no part in the reaction. In order to understand how this is decided, it is first necessary to outline some features of the structure of compounds.

**e** At AS you will only be required to write ionic equations for precipitations and the displacement reactions of halogens (see page 116). However, it will help you if you are able to provide a wider range of examples. Ionic equations are often easier to write and are accepted as an alternative if you are asked to give an equation for a reaction.

## Structure of compounds

Most compounds can be broadly classified as ionic or covalent and some useful generalisations can be made:

- Compounds containing metals are normally **ionic**. This means that they contain charged particles resulting from a loss or gain of electrons. The particles are called **ions**.
- Compounds containing only non-metals are **covalent**. This means that they contain neutral particles resulting from the sharing of electrons. The particles are called **molecules**.

However, there are some compounds that are naturally covalent but react with water and become ionic — for example, the mineral acids. Hydrogen chloride is a covalent gas, but when dissolved in water it becomes ionic. The ions $H^+(aq)$ and $Cl^-(aq)$ are formed. Similar behaviour occurs with nitric and sulphuric acids.

### Differences between ionic and covalent substances

Ionic bonds are formed by the electrical attraction between the positive and negative ions, but the ions are still separate from each other. When the compound is melted or dissolved in water the ions become detached from each other and move around freely and separately. This gives them the capacity to react independently.

Covalent substances have bonds that involve the sharing of electrons. The shared electrons bind the atoms together and the molecules stay as complete entities even when the substance is melted or dissolved.

Figure 2.1 illustrates the particles in an aqueous solution of sodium chloride, in which the sodium ions and chloride ions are moving freely and independently. In a solution of glucose — which is covalent — the glucose molecules would be intact (Figure 2.2).

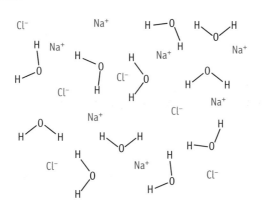

**Figure 2.1**
*An aqueous solution of sodium chloride. The ions, Na⁺ and Cl⁻, are separated and can react independently.*

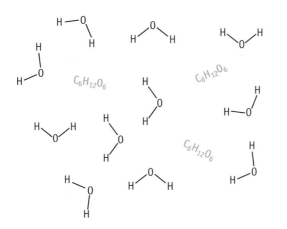

**Figure 2.2**
*An aqueous solution of glucose. The molecules of glucose, $C_6H_{12}O_6$, are intact.*

The distinction between the two types of bonding is important because:

- the reaction of an ionic substance in solution usually involves just one of the ions present in the substance
- the reaction of a covalent compound involves the whole molecule

For example, aqueous silver nitrate reacts with aqueous sodium chloride to produce insoluble silver chloride. However, it is more correct to say that silver ions from aqueous silver nitrate react with the chloride ions from aqueous sodium chloride to produce a precipitate of silver chloride. It does not matter that it is a nitrate of silver or the chloride of sodium — the reaction takes place as a consequence of silver ions and chloride ions being made mobile by the dissolving process. The reaction still occurs if the metal ion associated with the chloride is, for example, potassium, magnesium or copper and if the silver ion is dissolved as a sulphate. This reaction is the basis of a method for identifying a chloride in aqueous solution by adding an aqueous silver salt (pages 117–18).

A standard equation does not make this point clear. The reaction between silver nitrate and sodium chloride can be written as:

$$AgNO_3(aq) + NaCl(aq) \rightarrow AgCl(s) + NaNO_3(aq)$$

This provides useful information, particularly if state symbols are used. However, the equation does not show that only some of the ions are involved in the reaction. This is where ionic equations are helpful. The ionic equation is:

$$Ag^+(aq) + Cl^-(aq) \rightarrow AgCl(s)$$

It shows that only aqueous silver ions and aqueous chloride ions are involved in the reaction. Nitrate and sodium ions are present in the mixture but do not change. They move freely and independently at the start of the reaction and also when the reaction has finished. They are referred to as **spectator ions**. The presence of sodium nitrate as a product could be shown only by filtering the product mixture and evaporating the water. The ions would then combine and form a solid.

Ionic equations may seem strange because it is not possible to have silver or chloride ions in isolation; they have to be associated with other ions. However, an ionic equation is a truer representation of such reactions.

# Examples of ionic equations

Reaction between aqueous sodium hydroxide and hydrochloric acid

Aqueous sodium hydroxide reacts with hydrochloric acid to produce aqueous sodium chloride and water. This is represented by the equation:

$$NaOH(aq) + HCl(aq) \rightarrow NaCl(aq) + H_2O(l)$$

Water is covalent and is, therefore, present as intact covalent molecules. Sodium hydroxide and sodium chloride are ionic and, because HCl is in solution, so is the hydrochloric acid. The ions behave independently.

The sodium ions and chloride ions are not changed by the reaction — they remain in solution as free, independent ions. The hydroxide ions and the hydrogen ions react. They combine to form covalent water molecules.

This is shown by the ionic equation:

$$H^+(aq) + OH^-(aq) \rightarrow H_2O(l)$$

This ionic equation demonstrates the important fact that all hydroxides react with a source of hydrogen ions (i.e. an acid) to produce water.

Note that in the reaction between silver nitrate and sodium chloride, the product, silver chloride, is included in the ionic equation but here sodium chloride is absent. The difference is that silver chloride is formed as a solid with the silver ions and chloride ions held together in a crystal lattice, whereas sodium chloride is in solution, so the ions are free and independent.

Reaction between a carbonate and an acid

Aqueous sodium carbonate reacts with hydrochloric acid to produce sodium chloride, carbon dioxide and water:

$$Na_2CO_3(aq) + 2HCl(aq) \rightarrow 2NaCl(aq) + CO_2(g) + H_2O(l)$$

The sodium ions and chloride ions are not involved in the reaction — they remain unchanged in solution. The carbonate ions and hydrogen ions react to form the two covalent products, carbon dioxide and water. This is represented by the ionic equation:

$$CO_3{}^{2-}(aq) + 2H^+(aq) \rightarrow CO_2(g) + H_2O(l)$$

Ionic equations must be balanced. In the example above, a '2' has to be placed before the 'H+'. The equation indicates that all carbonates react with acids to form carbon dioxide and water. For sulphuric acid, the equation is:

$$Na_2CO_3(aq) + H_2SO_4(aq) \rightarrow Na_2SO_4(aq) + CO_2(g) + H_2O(l)$$

The ionic equation is the same as for the reaction between sodium carbonate and hydrochloric acid:

$$CO_3{}^{2-}(aq) + 2H^+(aq) \rightarrow CO_2(g) + H_2O(l)$$

**e** Do not be tempted to write '$H_2{}^+(aq)$' or '$H_2{}^{2+}(aq)$' to represent the hydrogen ions arising from $H_2SO_4$. Once in solution, the ions are detached and are simply $H^+$. The equation shows that two hydrogen ions are needed for each carbonate ion. '$H_2{}^+(aq)$' implies that the two hydrogen atoms are bonded covalently and have acquired a positive charge; '$H_2{}^{2+}(aq)$' indicates the same, but with a charge of '2+'. Neither representation is correct.

Reaction between zinc and aqueous copper sulphate

Zinc metal reacts with aqueous copper sulphate to form copper metal and aqueous zinc sulphate, according to the equation:

$$Zn(s) + CuSO_4(aq) \rightarrow Cu(s) + ZnSO_4(aq)$$

Copper sulphate and zinc sulphate are both ionic, so the ionic equation for this reaction is:

$$Zn(s) + Cu^{2+}(aq) \rightarrow Cu(s) + Zn^{2+}(aq)$$

Reaction between solid magnesium oxide and hydrochloric acid

Remember that ionic substances must be in solution to exist as free ions. The equation for the reaction of solid magnesium oxide with hydrochloric acid to form magnesium chloride and water is as follows:

$$MgO(s) + 2HCl(aq) \rightarrow MgCl_2(aq) + H_2O(l)$$

In this case, the ions in the solid magnesium oxide are held in an ionic lattice and are not free to move. The only ions that are 'free' throughout this reaction are the chloride ions. Therefore, the ionic equation is:

$$MgO(s) + 2H^+(aq) \rightarrow Mg^{2+}(aq) + H_2O(l)$$

During this reaction the magnesium oxide lattice breaks down, releasing magnesium ions into the solution. The oxide ions combine with the hydrogen ions from the acid to form water.

## Questions

**1** Identify which of the following substances are ionic and give the formulae of the ions formed when the substances are dissolved in an aqueous solution.

**a** $NaNO_3$

**b** $NO$

**c** $CCl_4$

**d** $Pb(NO_3)_2$

**e** $K_2SO_4$

**f** $CH_4$

**g** $H_2SO_4$

**h** $MgCl_2$

**2** Write ionic equations for the following reactions. Include appropriate state symbols.

**a** silver nitrate solution + potassium iodide solution →
silver iodide precipitate + potassium nitrate solution

**b** potassium hydroxide solution + nitric acid →
potassium nitrate solution + water

**c** zinc metal + hydrochloric acid → zinc chloride solution + hydrogen gas

**d** magnesium metal + copper sulphate solution →
magnesium sulphate solution + copper metal

**e** lithium carbonate solution + hydrochloric acid →
lithium chloride solution + carbon dioxide + water

**f** sodium hydroxide solution + copper nitrate solution →
copper hydroxide precipitate + sodium nitrate solution

**g** sodium hydroxide solution + aluminium sulphate solution →
aluminium hydroxide precipitate + sodium sulphate solution

**h** zinc oxide solid + hydrochloric acid → zinc chloride solution + water

# Summary

You should now be able to:

- deduce the formula of a compound
- write balanced equations
- write ionic equations

# Additional reading

Symbols for the elements were first suggested by Jöns Jacob Berzelius in 1811:

> I must observe here that the object of the new signs is not that…they should be employed to label vessels in the laboratory: they are destined solely to facilitate the expression of chemical proportions, and to enable us to indicate, without long periphrases, the relative number of volumes of the different constituents contained in each compound body.

He also felt that 'the chemical signs ought to be letters, for the greater facility of writing, and not to disfigure a printed book'. Most of Berzelius's symbols are still used today.

However, in the nineteenth century, representing compounds with symbols was not straightforward. Even for simple compounds, it required ingenuity and effort.

Take water as an example. It can be shown by experiment that 1g of hydrogen reacts exactly with 8g of oxygen. This could be because the formula of water is HO, with an atom of oxygen having a mass eight times greater than that of hydrogen. Equally, the formula could be $H_2O$, with the oxygen atom having a mass 16 times greater than that of hydrogen. It may be that the formula is $HO_2$, with an atom of oxygen having a mass four times greater than that of hydrogen. Although the answer is known now, it could not be deduced from these experimental results alone.

Scientists had to guess and then try to make sense of other results. Suppose they thought that the formula of water was HO (as did John Dalton). This makes oxygen have a mass of 8 compared with that of hydrogen. However, when sulphur and oxygen react to form sulphur dioxide, 8g of oxygen reacts with 8g of sulphur. Two atoms cannot have the same mass, so a decision is required. It might be decided that the formula of sulphur dioxide is $S_2O$, with sulphur having a mass four times that of hydrogen. However, when hydrogen and sulphur react, 1g of hydrogen reacts with 16g of sulphur. This indicates that the formula of hydrogen sulphide should be $HS_4$. Scientists would realise that this is unlikely, since they suppose that the ratio of atoms in compounds should be simple. This might prompt a re-think of the assumptions made. Eventually, by guesswork and the elimination of various possibilities, the right answer might be found — or at least one that seems consistent. This was the process that had to be used originally to establish correct formulae.

Gradually, a consistent scheme was developed, using ideas suggested by Amedeo Avogadro, who did extensive work on the reacting quantities of gases (Chapter 5). It is interesting though that, even by the 1860s, there was still some uncertainty about the correct mass of a carbon atom relative to a hydrogen atom and consequently some uncertainty over the formulae of carbon compounds.

## Extension questions: blue stars, pigments and better beer

1 One of the biggest challenges for firework manufacturers is creating satisfactory blue colours for stars and cascades. Copper chloride can give this colour when mixed with a reagent such as potassium chlorate, but it tends to be unstable at high temperature and the colour is then lost.

Copper chloride is manufactured either by the reaction of chlorine directly with copper metal or by dissolving copper oxide in dilute hydrochloric acid. The potassium chlorate ($KClO_3$) releases oxygen to the firework mixture by decomposing to form potassium chloride.

a Write an equation to represent:
  (i) the reaction between chlorine and copper metal
  (ii) the reaction between copper oxide and dilute hydrochloric acid
  (iii) the decomposition of potassium chlorate

A mixture that has been proposed to intensify the blue colour includes a pigment called Paris Green. This pigment is a salt containing one part of copper ethanoate and three parts of copper arsenite. The ethanoate ion has the formula $CH_3CO_2^-$. The formula of the arsenite ion is $AsO_2^-$.

b Write the formula for:
(i) copper ethanoate
(ii) copper arsenite

2 Scheele's Green (copper hydrogenarsenate, $CuHAsO_3$) was developed in the nineteenth century to improve the variety and quality of dyes. It was used by wallpaper manufacturers; William Morris included it as a dye. They were either unaware of or indifferent to the fact that the formulation is poisonous. Miners at the Devon Great Consuls mine near Tavistock who extracted the arsenic ore from which Scheele's Green was made suffered badly from the effects of their work.

A recipe from the nineteenth century describes a process for making Scheele's Green:

■ A mixture of arsenic oxide is boiled with potassium carbonate until no more carbon dioxide is produced. In this reaction potassium hydrogenarsenate is also formed.

■ A solution of copper sulphate is then added to this mixture and an impure precipitate of Scheele's Green is formed.

Some purification is then required before a usable powder is obtained.

a Deduce the formula of potassium hydrogenarsenate.

b Suggest an equation for the reaction described in the first bullet point above. (Note: water is required as a reactant.)

c Suggest an equation for the reaction in the second bullet point.

3 A surprising use of copper is to remove hydrogen sulphide from beer. During fermentation, hydrogen sulphide is produced as a natural by-product. However, because of its unpleasant odour, even tiny amounts can spoil the flavour of the beer. Low concentrations of copper ions are sufficient to give a precipitate of copper sulphide, which is then removed, leaving the beer free from this impurity.

Write an ionic equation for this reaction.

# Relative atomic mass and the mole

## Chapter 3

This chapter examines how equations and atomic masses are used to calculate the quantities required for chemical reactions. There are several definitions involved and the minor distinctions between the terms used may at first seem confusing. However, to understand how quantitative chemistry works, you must learn the precise meaning of each term.

The term 'mole' — the amount of substance — is particularly important because it is the basis for the calculations covered in this and subsequent chapters.

## Relative atomic mass

Nineteenth-century chemists established the formulae of common substances and also calculated the relative masses of the atoms from which they are made. It was by no means an easy task and it took several decades before agreement was reached. What they discovered was not the individual masses of the atoms — they knew that atoms were small and would have assumed that it would never be possible to know these values — but how the masses of atoms compared with each other. For example, it was established that a sulphur atom was twice as heavy as an oxygen atom and that an oxygen atom was 16 times heavier than a hydrogen atom. A nominal value of 1 was assigned to the mass of the lightest atom, hydrogen, and the masses of other elements were reported relative to this standard. Thus, the mass of an oxygen atom is approximately 16 and that of a sulphur atom, 32. Every element can be given a number that indicates its atomic mass relative to that of a hydrogen atom. These numbers relate to the sum of the numbers of protons and neutrons present in the nucleus of an atom (Chapter 1).

For many years this was the basis on which the masses of atoms were compared. The choice of the lightest element as the standard may seem obvious, but the situation becomes more complex when isotopes are taken into account. This is one reason why the standard element used for comparison today is the isotope of carbon that has mass number 12.

The mass of an element compared to carbon-12 is called the **relative atomic mass** and is defined as follows:

> Relative atomic mass is the weighted mean mass of an atom of an element compared with $\frac{1}{12}$ the mass of an atom of carbon-12, which is taken as 12 exactly.

The expression 'weighted mean mass' takes into account isotopes that are present in different amounts. For example, chlorine is a mixture of 75% chlorine-35 atoms and 25% chlorine-37 atoms. Its relative atomic mass is averaged as $(0.75 \times 35) + (0.25 \times 37)$, which is equal to 35.5.

Relative atomic mass has no units because it is a comparison between the element and carbon-12.

Relative atomic masses are listed in most versions of the periodic table (page 264). They will be available to you in an examination, so you do not have to learn them.

## Definitions linked to relative atomic mass

Other definitions follow from the definition for relative atomic mass. For a particular isotope of an element, the definition is slightly modified:

Relative isotopic mass is the mass of an atom of an isotope of the element compared with $\frac{1}{12}$ the mass of an atom of carbon-12, which is taken as 12 exactly.

A similar statement can be made for compounds. If the compound is covalent (Chapter 6), the smallest unit is a molecule, so the term **relative molecular mass** is used:

Relative molecular mass is the weighted mean mass of a molecule of the compound compared with $\frac{1}{12}$ the mass of an atom of carbon-12, which is taken as 12 exactly.

For ionic substances (e.g. those containing a metal) the smallest units are the ions indicated by their formulae and the expression **relative formula mass** should be used:

Relative formula mass is the weighted mean mass of a formula unit of the compound compared with $\frac{1}{12}$ the mass of an atom of carbon-12, which is taken as 12 exactly.

Relative formula mass can also be used when referring to covalent compounds.

Obtaining relative molecular mass and relative formula mass

The relative molecular or formula mass of a compound is obtained by adding together the relative atomic masses of its component atoms. In each of the following examples, the values of the relative atomic masses have been obtained from the periodic table (page 264).

> *Worked examples*
> 1 Methane has the formula $CH_4$.
>    Its relative molecular mass is $12 + (4 \times 1) = 16$.
> 2 Glucose has the formula $C_6H_{12}O_6$.
>    Its relative molecular mass is $(6 \times 12) + (12 \times 1) + (6 \times 16) = 180$.
> 3 Magnesium carbonate has the formula $MgCO_3$.
>    Its relative formula mass is $24.3 + 12.0 + (3 \times 16.0) = 84.3$.

*e* It is a common mistake in exams to give relative atomic mass a unit. Remember that it is just a number with no units.

*e* When using the periodic table, make sure that you do not use the atomic number by mistake. For example, the atomic number of oxygen is 8, but its relative atomic mass is 16.

*e* Relative formula mass can be helpful if you are uncertain whether the substance is ionic or covalent.

**4** Calcium hydroxide has the formula $Ca(OH)_2$.
Its relative formula mass is $40.1 + 2 \times (16.0 + 1.0) = 74.1$.

**5** Copper sulphate crystals have the formula $CuSO_4.5H_2O$. This means that there are five molecules of water for each $CuSO_4$ unit.
Its relative formula mass is $63.5 + 32.1 + (4 \times 16.0) + (5 \times 18.0) = 249.6$.

Remember that $(OH)_2$ means that there are two OH units for every calcium, so the relative atomic masses of both oxygen and hydrogen must be doubled.

Note that 'relative formula mass' is used in examples (3)–(5) instead of relative molecular mass, because magnesium carbonate, calcium hydroxide and copper sulphate are ionic (since they contain metals).

With more complicated formulae, be careful to obtain the correct number of atoms of each type. For example, calcium nitrate has the formula $Ca(NO_3)_2$, which means that it contains one calcium, two nitrogens and six oxygens — the brackets mean that there are two $NO_3^-$ ions.

## Questions

**1** For each of the following, write the formula and calculate the relative formula mass:
  **a** Lithium chloride
  **b** Potassium bromide
  **c** Magnesium hydroxide
  **d** Sulphuric acid
  **e** Sodium sulphate
  **f** Barium nitrate
  **g** Iron(II) chloride
  **h** Iron(III) sulphate
  **i** Sodium carbonate crystals ($Na_2CO_3.10H_2O$)
  **j** Iron(II) sulphate crystals ($FeSO_4.7H_2O$)

**2** Calculate the relative molecular mass of each of the following covalent compounds:
  **a** Ethane, $C_2H_6$
  **b** Ethanol, $C_2H_5OH$
  **c** Ethanoic acid, $CH_3COOH$
  **d** Chloromethane, $CH_3Cl$
  **e** Aminoethane, $C_2H_5NH_2$
  **f** Nitrogen gas

# The mole

To prepare a cake, the quantities of ingredients have to be measured. Solids, such as flour, are weighed; liquids are measured by volume. Laboratories have more accurate measuring devices — for example, balances to record masses to four decimal places and volumes measured to two decimal places. These are necessary because the preparation of chemicals requires more precision than that required for making a cake.

If laboratory measurements are more accurate, you might conclude that assembling the correct quantities for a reaction would be straightforward, and this is sometimes true. However, it is important to understand that the information available about the quantities required for a chemical reaction can only be obtained from a balanced equation for the reaction. This is why balancing equations is such an important skill.

## Questions

For the following balanced equations, write down the information that each provides about the number of atoms or molecules that react and that are produced.

a $Mg + S \rightarrow MgS$

b $S + O_2 \rightarrow SO_2$

c $Zn + I_2 \rightarrow ZnI_2$

d $N_2 + O_2 \rightarrow 2NO$

e $N_2 + 3H_2 \rightarrow 2NH_3$

f $2Al + 3O_2 \rightarrow Al_2O_3$

## The Avogadro constant

Although an equation gives precise information about the numbers of particles (atoms, molecules etc.) in a reaction, these particles can be neither seen nor counted. In a laboratory, substances can only be weighed or their volumes measured.

However, if one atom of zinc requires one atom of sulphur to make one unit of zinc sulphide, then 1000 atoms of zinc need 1000 atoms of sulphur to make 1000 units of zinc sulphide. By increasing the number of particles, keeping the ratio of one atom of sulphur to one atom of zinc, an amount substantial enough to be weighed is eventually reached.

The scientists who painstakingly compiled the relative masses of the atoms in the nineteenth century might not have thought it possible to count atoms and use this information to deduce the correct mass needed for a reaction. However, advances in understanding and the availability of sophisticated measuring devices have made this possible.

It turns out that $6.02 \times 10^{23}$ atoms ($602\,000\,000\,000\,000\,000\,000\,000$ atoms) of carbon are needed to obtain a mass of 12 g. Since 12 g of carbon is the established standard, the value $6.02 \times 10^{23}$ atoms is an important quantity known as the **Avogadro constant**.

> The mass of an element or compound that contains $6.02 \times 10^{23}$ particles is known as the mole.

> The symbol for the Avogadro constant is $N_A$.

## Calculations

One mole is an amount of a substance that contains a fixed number of particles (the Avogadro constant). Therefore, in the reaction between zinc and sulphur, one mole of zinc reacts with one mole of sulphur and one mole of zinc sulphide is produced. By using one mole of each substance the ratio of one atom of sulphur

It has been estimated that if Julius Caesar's dying breath were evenly distributed in the atmophere, then every time we breathe in, each person would inhale 100 molecules of it. This emphasises the fact that molecules and atoms are extremely small particles.

The Avogadro constant used to have the symbol 'L' because it was known as the Loschmidt number.

The abbreviation of mole is 'mol'. Since it is a mass, it has units.

to one atom of zinc is maintained. The mass of a mole of any element can be found using the periodic table (page 264). In this case, zinc is given as 65.4 g and sulphur as 32.1 g, i.e. the relative atomic mass plus the unit of grams. Therefore, 65.4 g of zinc reacts with 32.1 g of sulphur to form 97.5 g of zinc sulphide.

Using mass to count the correct number of particles is not unique to chemistry. If a bank customer asked for 1000 20p coins, these would be weighed, not counted. It is known that 1000 20p coins weigh exactly 5 kg. If a 'bank mass' is defined as 1000 coins, a table of masses for 1000 coins could be created. For example, a 'bank mass' of £1 coins is 9.500 kg and of 10p coins is 6.500 kg.

Consider the reaction between iron and sulphur:

$$2Fe + 3S \rightarrow Fe_2S_3$$

This equation shows that two atoms of iron require three atoms of sulphur. This ratio is maintained by taking two moles of iron and reacting them with three moles of sulphur.

One mole of iron has a mass of 55.8 g and one mole of sulphur has a mass of 32.1 g. Therefore, 111.6 g ($2 \times 55.8$) of iron reacts with 96.3 g ($3 \times 32.1$) of sulphur. Mass cannot be lost when a chemical reaction takes place, so the mass of iron(III) sulphide formed is $111.6 + 96.3 = 207.9$ g.

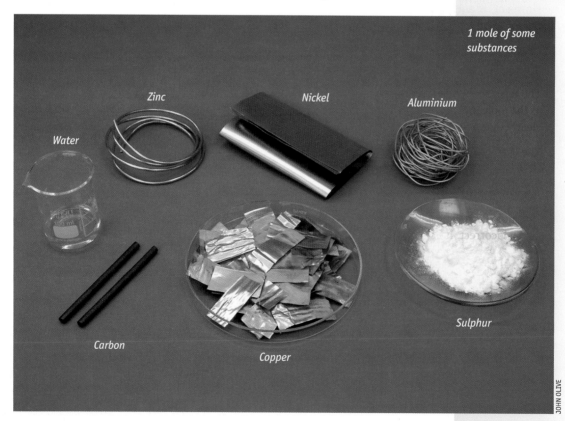

1 mole of some substances

Water

Zinc

Nickel

Aluminium

Carbon

Copper

Sulphur

JOHN OLIVE

Further calculations are given on pages 47–48. Note that the mass of one mole of a compound is obtained by adding together the masses of each of its components. The procedure is the same as that used for relative molecular (or formula) mass, but the result obtained is a mass in grams, rather than a number with no units.

> **Worked example**
>
> What is the mass of 1 mol of potassium oxide?
>
> **Answer**
>
> The formula of potassium oxide is $K_2O$.
>
> 1 mol of potassium, K, has a mass of 39.1 g; 1 mol of oxygen has a mass of 16.0 g. 1 mol of potassium oxide, $K_2O$, contains 2 mol of potassium, K, and 1 mol of oxygen, O.
>
> So the mass of 1 mol of $K_2O$ is $(2 \times 39.1) + (1 \times 16.0) = 94.2$ g.

## Question

Calculate the mass of one mole of the following compounds:

a Carbon dioxide
b Ethane ($C_2H_6$)
c Zinc oxide
d Zinc chloride
e Copper carbonate
f Nitric acid

## Molar mass

The term **molar mass** is often used. Molar mass has the same numerical value as the mole, but it has the units $g\,mol^{-1}$. This is helpful because it indicates more clearly that a mole of an element or compound is a mass.

The correct definition of a mole is as follows:

> The mole is the mass of an element or compound that contains exactly the same number of particles as there are atoms in 12 g of carbon-12.

The mole used to be thought of as a number equal to the Avogadro constant, but this is incorrect. It is an amount of substance that usually has units of grams.

# Summary

Relative (atomic, isotopic, molecular or formula) masses show how the masses of elements and compounds compare with each other, based on a scale in which carbon-12 is given a value of 12. Since they are comparative, they do not have units.

A mole is an amount equal to the number of particles in 12 g of carbon-12. The number of particles is $6.02 \times 10^{23}$, the Avogadro constant. The mole has units of mass.

To emphasise how the terms are related and to show how careful you must be to supply the correct units, note the following:

- The relative molecular mass of methane, $CH_4$, is 16.0.
- The mass of 1 mol of methane is 16.0 g.
- The molar mass of methane is 16.0 g mol$^{-1}$.
- The relative formula mass of zinc sulphide is 97.5.
- The mass of 1 mol of zinc sulphide is 97.5 g.
- The molar mass of zinc sulphide is 97.5 g mol$^{-1}$.

You should now be able to:

- define relative atomic mass and relative isotopic mass
- calculate relative molecular mass and relative formula mass
- calculate relative atomic mass using percentage isotopic masses
- define the mole

# Additional reading

The use of carbon-12 ($^{12}C$) as the standard for measuring relative atomic masses was only agreed in 1961. Dalton had suggested that, since hydrogen was the lightest element, setting its value as 1 was logical. However, in the nineteenth century, atomic masses were determined by reacting elements together and establishing the proportions that reacted. Hydrogen, therefore, was no longer felt to be a good choice because it does not react directly with many other elements. The Swedish chemist Berzelius suggested oxygen as the standard and gave it an arbitrary value of 100. Others agreed with the use of oxygen but proposed values of less than 100.

This resulted in chaos, which was resolved eventually by the choice of 16 for oxygen. This had the merit of establishing a link with Dalton's use of 1 for hydrogen. However, it was not until 1903 that a formal agreement was reached. This still did not resolve the matter. In general, physicists continued to use oxygen-16 as the standard. However, as more isotopes were discovered, chemists felt it would be better use carbon-12. To the first decimal place, this makes little difference to the values of relative atomic masses. However, these values were beginning to be calculated to four decimal places and, at this precision, the standard does matter. In 1961, an international agreement to use carbon-12 was reached.

The number of atoms contained in 1 mol of carbon-12 is now known universally as the Avogadro constant, although Avogadro made no attempt to calculate its value. The first calculation was carried out by an Austrian schoolteacher called Josef Loschmidt and, in Germany, the value was known for many years as the Loschmidt number and given the symbol $L$. It was first called the Avogadro constant in a paper written by Jean Perrin in 1909.

The number is not easy to calculate and many attempts were made in the twentieth century. By the

## Amedeo Avogadro, 1776–1856

Lorenzo Romano Amedeo Carlo Avogadro was born into a wealthy family in Turin. Initially, he studied law and received a doctorate in ecclesiastical law at the age of 20. He practised law for 3 years and occasionally returned to it during his lifetime. His real interest was science (natural philosophy, as it was then called) and he made crucial contributions to our understanding of atoms and molecules. He was married and had six sons. In 1787, he inherited his father's title of Count of Quaregna. He was a professor of mathematical physics at Turin for 25 years and held a range of public offices concerned with national statistics, meteorology and weights and measures. He introduced the metric system to his native Piedmont.

SPL

early 1930s, there were as many as 80 different values, which may reflect the different standard values being used for relative atomic masses. They were all close to the currently accepted value of $6.02214199 \times 10^{23}$.

One method for determining the Avogadro constant uses the structure of metals. In metals, atoms are arranged in well-defined structures. For example, the elements in group 1 of the periodic table have a repeating pattern of a metal atom at the centre of a cube surrounded by eight metal atoms on the corners of the cube. The size of the cube can be determined using X-rays. This allows the volume, $v$, occupied by one atom of the metal to be calculated. Using the relative atomic mass of the element and its density, the volume, $V_m$, of 1 mol can be established, since volume equals mass/density. The Avogadro number is then determined by $V_m/v$.

## Extension questions: the mole

1 Consider sugar lumps to be cubes of side 1 cm. Suppose that 1 mol of sugar lumps were laid on all the land on the Earth's surface. The Earth's land surface is approximately $150\,000\,000\,km^2$.

a Calculate the area of the Earth's land surface in $cm^2$.

b Work out how many sugar lumps would be required.

c Taking Avogadro's number to be $6 \times 10^{23}$, calculate the height of the pile of sugar lumps in metres.

2 a For an element, the mole is the weighted mean mass of all its isotopes. Individually, a mole of each isotope has a different value. For example, to two significant figures, a mole of chlorine as $^{35}Cl$ has a mass of 35 g and a mole of $^{37}Cl$ has a mass of 37 g. These are the values that would be obtained using a mass spectrometer. The $^{35}Cl$ isotope has a relative abundance of 75% and $^{37}Cl$ has a relative abundance of 25%.

Calculate the mass of 1 mol of chlorine atoms.

b Chlorine occurs naturally as $Cl_2$. The mass of 1 mol of this molecule is made up of contributions from all the variations in mass of the isotopes.

  (i) Explain why $Cl_2$ molecules have masses of 70, 72 and 74.

  (ii) Explain which $Cl_2$ molecule would be most abundant.

  (iii) Calculate the relative abundance of each $Cl_2$ molecule.

  (Part (iii) is for students studying mathematics.)

3 A hydrogen atom consists of 1 proton and 1 electron. The mass of:

■ a proton is $1.672648586 \times 10^{-27}$ kg

■ an electron is $0.910953448 \times 10^{-30}$ kg

Calculate the mass of 1 mol of hydrogen atoms. Give the answer in grams.

# Mole calculations

## Chapter 4

In this chapter, the masses of chemical substances required for, and produced in, chemical reactions are considered. Some reactions involve the evolution of a gas, and the relationship between the volume of a gas and the amount in moles is established.

You should have acquired the necessary background chemical understanding from Chapter 3; this chapter is mainly concerned with putting those ideas into practice. Make sure that you are comfortable with the arithmetic involved. Some students find this difficult at first and you may need to do plenty of examples before the necessary fluency is obtained. Once this has been achieved, you will be able to focus confidently on the chemistry.

## Converting masses to moles and moles to masses

In a laboratory, quantities are measured by the mass of a substance or the volume of a liquid. An equation, however, provides information about the numbers of particles. This can also, and more helpfully, be expressed as the amount in moles. Therefore, the mass needs to be related to the amount in moles indicated by the equation. The relationship between the number of grams and the amount in moles requires the use of molar mass, which has units of $g\,mol^{-1}$.

> **Worked example 1**
> Calculate the amount in moles of 6 g of carbon.
>
> **Answer**
> The molar mass of carbon is $12.0\,g\,mol^{-1}$. This means that 1 mol has a mass of 12.0 g.
>
> $$\text{mass of 6 g expressed as an amount in moles} = \frac{6\,g}{12.0\,g\,mol^{-1}} = 0.5\,mol$$

In worked example 1, it can be seen immediately that 6 g is 0.5 mol because the numbers are simple. You should nevertheless appreciate that what you have done is to use the formula:

$$\text{amount in moles } (n) = \frac{\text{mass in grams } (m)}{\text{molar mass } (M)}$$

### Worked example 2
Calculate the amount in moles of 4.52 g of carbon.

**Answer**

molar mass of carbon is $12.0\,\text{g mol}^{-1}$

$$\text{amount in moles} = \frac{\text{mass in grams}}{\text{molar mass}} = \frac{4.52\,\text{g}}{12.0\,\text{g mol}^{-1}} = 0.377\,\text{mol}$$

### Worked example 3
Calculate the amount in moles of 5.0 g of chlorine gas.

**Answer**

molar mass of $Cl_2 = 71.0\,\text{g mol}^{-1}$

$$\text{amount in moles} = \frac{\text{mass in grams}}{\text{molar mass}} = \frac{5.0\,\text{g}}{71.0\,\text{g mol}^{-1}} = 0.070\,\text{mol}$$

◀ Remember that chlorine gas has the formula $Cl_2$.

### Worked example 4
Calculate the amount in moles of 8.48 g of calcium sulphate.

**Answer**

molar mass of calcium sulphate, $CaSO_4 = 40.1 + 32.1 + (4 \times 16.0) = 136.2\,\text{g mol}^{-1}$

$$\text{amount in moles} = \frac{\text{mass in grams}}{\text{molar mass}} = \frac{8.48\,\text{g}}{136.2\,\text{g mol}^{-1}} = 0.0623\,\text{mol}$$

### Worked example 5
Calculate the mass (in grams) of 0.25 mol of calcium carbonate.

**Answer**

molar mass of calcium carbonate, $CaCO_3 = 40.1 + 12.0 + (3 \times 16.0) = 100.1\,\text{g mol}^{-1}$
mass of 0.25 mol = $0.25\,\text{mol} \times 100.1\,\text{g mol}^{-1} = 25.025\,\text{g}$

◀ You may have spotted that 0.25 is $\frac{1}{4}$, so the mass must be $\frac{1}{4} \times 100.1\,\text{g}$.

The formula used in worked example 5 is as follows:

mass in grams ($m$) = amount in moles ($n$) × molar mass ($M$)

### Worked example 6
Calculate the mass (in grams) of 0.107 mol of sulphuric acid.

**Answer**

molar mass of sulphuric acid, $H_2SO_4 = (2 \times 1.0) + 32.1 + (4 \times 16.0) = 98.1\,\text{g mol}^{-1}$
mass in grams of 0.107 mol = $0.107\,\text{mol} \times 98.1\,\text{g mol}^{-1} = 10.5\,\text{g}$

# Calculating molar masses

If the amount in moles and the mass in grams are both known, the molar mass of an element or compound can be calculated. The relationship is:

$$\text{molar mass} = \frac{\text{mass in grams}}{\text{amount in moles}}$$

---

**Worked example 1**

Calculate the molar mass of an element if 0.10 mol has a mass of 6.4 g.

**Answer**

$$\text{molar mass} = \frac{\text{mass in grams}}{\text{amount in moles}} = \frac{6.4\,g}{0.10\,mol} = 64\,g\,mol^{-1}$$

---

**Worked example 2**

Calculate the molar mass of an acid if 0.071 mol has a mass of 6.96 g.

**Answer**

$$\text{molar mass} = \frac{\text{mass in grams}}{\text{amount in moles}} = \frac{6.96\,g}{0.071\,mol} = 98\,g\,mol^{-1}$$

---

The following triangle may help you remember how to tackle these calculations.

To determine molar mass ($M$), cover that corner of the triangle. This shows that molar mass = mass ÷ amount in moles, or $M = m/n$.

To determine mass, cover the $m$ corner of the triangle, which gives $m = n \times M$.

To determine moles, cover the $n$ corner of the triangle, which gives $n = m/M$.

## Questions

1 Determine the amount in moles in each of the following:
   a 8.0 g of sulphur
   b 2.4 g of magnesium
   c 4.8 g of oxygen
   d 1.68 g of calcium oxide
   e 1.0 g of calcium carbonate
   f 6.0 g of silver nitrate
   g 1.3 g of sodium sulphate
   h 5.0 g of aluminium sulphate

**2** Calculate the mass (in grams) of each of the following:
  **a** 0.5 mol of argon
  **b** 1.5 mol of sulphur
  **c** 0.25 mol of calcium
  **d** 0.2 mol of potassium chloride
  **e** 0.6 mol of magnesium sulphate
  **f** 5 mol of hydrogen gas
  **g** 0.04 mol of aluminium chloride
  **h** 0.45 mol of aluminium hydroxide

**3** Calculate the molar mass of each of the following and identify the element:
  **a** 0.1 mol of an element that has a mass of 10.80 g
  **b** 0.05 mol of an element that has a mass of 2.60 g
  **c** 0.55 mol of an element that has a mass of 12.65 g
  **d** 1.3 mol of an element that has a mass of 58.5 g

**4** Calculate the molar mass of each of the following compounds in which:
  **a** 0.03 mol has a mass of 1.92 g
  **b** 0.75 mol has a mass of 43.88 g
  **c** 0.35 mol has a mass of 64.40 g
  **d** 0.02 mol has a mass of 2.47 g
  **e** 0.91 mol has a mass of 212 g

**5** A compound contains a divalent metal, X. 0.02 mol of the hydroxide of X is found to have a mass of 1.48 g. Identify the metal, X.

# Calculating reacting masses from equations

Equations provide information that can be used, with molar masses, to determine the correct masses of reagents for a chemical reaction. However, we often work with masses other than the molar masses.

## How much is needed? How much is made?

*Zinc and sulphur*

Consider the reaction between zinc and sulphur. Suppose we want to react 4.88 g of zinc with the correct mass of sulphur to convert it completely into zinc sulphide and also to calculate the expected mass of zinc sulphide product.

The equation for the reaction is:

$$Zn + S \rightarrow ZnS$$

This shows that 1 mol of zinc requires 1 mol of sulphur — the mole ratio is 1:1.

First, establish the amount of zinc by calculating the amount in moles represented by 4.88 g. As 1 mol of zinc has a mass of 65.4 g, in 4.88 g the amount in moles is 4.88/65.4, which is 0.0746 mol.

Therefore, 4.88 g of zinc is the same as 0.0746 mol. This means that, to keep the mole ratio of zinc to sulphur as 1:1, 0.0746 mol of sulphur is needed.

1 mol of sulphur has a mass of 32.1 g. Only 0.0746 mol is required, so the mass needed is $0.0746 \times 32.1 = 2.39$ g.

The equation also shows that 1 mol of ZnS is made from 1 mol of Zn reacting with 1 mol of S. Hence, using 0.0746 mol of Zn should produce 0.0746 mol of ZnS. The molar mass of ZnS is $65.4 + 32.1 = 97.5$ g. Hence the mass produced is $0.0746 \times 97.5 = 7.27$ g.

An alternative method for carrying out this calculation is shown in Figure 4.1.

|  | Zn | + | S | → | ZnS |
|---|---|---|---|---|---|
| Mole ratio is | 1 | : | 1 | : | 1 |
| Mass of Zn | 4.88 |  |  |  |  |
| Convert to moles | $\dfrac{4.88}{65.4}$ = 0.0746 |  |  |  |  |

The benefit of the balanced equation is that once we know the number of moles of one reactant we know all the others

| 0.0746 | 0.0746 | 0.0746 |
|---|---|---|
|  | These can now both be converted to mass |  |
|  | $0.0746 \times 32.1$ = 2.39 g | $0.0764 \times 97.5$ = 7.27 g |

Check that the mass of the reactants equals the mass of the products, i.e. that matter has not been created or destroyed

| mass of Zn | + | mass of S | = | mass of ZnS |
|---|---|---|---|---|
| 4.88 g | + | 2.39 g | = | 7.27 g |

*Figure 4.1 Mole calculations for the reaction between zinc and sulphur*

### Iron and sulphur

Consider the reaction between iron and sulphur. What mass of sulphur is required to react with 5.58 g of iron to make iron(III) sulphide?

The equation is:

$$2Fe + 3S \rightarrow Fe_2S_3$$

Even more care is needed here because the equation indicates that 2 mol of iron react with 3 mol of sulphur.

From the equation, as 1 mol of iron has a mass of 55.8 g, the amount in moles is $5.58/55.8 = 0.1$ mol.

1 mol of iron requires 1.5 mol of sulphur and produces 0.5 mol of $Fe_2S_3$. Hence, 0.1 mol of Fe needs $0.1 \times 1.5$ mol of sulphur = 0.15 mol, which together produce 0.05 mol of iron(III) sulphide.

Therefore, the mass of sulphur required is $0.15 \times 32.1 = 4.82$ g, and 10.40 g of $Fe_2S_3$ is produced.

An alternative method for carrying out the calculation is shown in Figure 4.2.

| | 2Fe | + | 3S | → | Fe₂S₃ |

|  | 2Fe |  | 3S |  | Fe₂S₃ |
|---|---|---|---|---|---|
| Mole ratio is | 2 | : | 3 | : | 1 |
| Mass of Fe | 5.58 | | | | |
| Convert to moles | $\frac{5.58}{55.8}$ = 0.1 | | | | |

The benefit of the balanced equation is that once we know the number of moles of one reactant we know all the others

| 0.1 | 0.15 | 0.05 |
|---|---|---|
| | These can now both be converted to mass | |
| | $0.15 \times 32.1$ = 4.82 g | $0.05 \times 207.9$ = 10.40 g |

Check that the mass of the reactants equals the mass of the products, i.e. that matter has not been created or destroyed

| mass of Fe | + | mass of S | = | mass of Fe₂S₃ |
|---|---|---|---|---|
| 5.58 g | + | 4.82 g | = | 10.40 g |

*Figure 4.2 Mole calculations for the reaction between iron and sulphur*

ℯ It is essential to your study of chemistry that you master such calculations. Work through the examples carefully and make sure that you understand them.

---

**Worked example 1**

What mass of zinc carbonate produces 4.20 g of zinc oxide when heated?

$$ZnCO_3 \rightarrow ZnO + CO_2$$

**Answer**

The equation indicates that 1 mol of zinc carbonate produces 1 mol of zinc oxide. 1 mol of zinc oxide has a mass of $65.4 + 16.0 = 81.4$ g.

4.20 g is $\frac{4.20}{81.4}$ mol = 0.0516 mol

Therefore, 0.0516 mol of zinc carbonate is needed.

1 mol of zinc carbonate has a mass of $65.4 + 12.0 + (3 \times 16.0) = 125.4$ g

mass of 0.0516 mol of zinc carbonate is $0.0516 \times 125.4 = 6.47$ g

Therefore, 6.47 g of zinc carbonate is required.

### Worked example 2

What mass of aluminium oxide can be obtained by reacting 8.00 g of aluminium with oxygen?

$$4Al + 3O_2 \rightarrow 2Al_2O_3$$

**Answer**

The equation indicates that 4 mol of aluminium react with 3 mol of oxygen to produce 2 mol of aluminium oxide.

1 mol of aluminium has mass 27.0 g.

$$8.00\,g\ aluminium\ is\ \frac{8.00}{27.0}\,mol = 0.30\,mol$$

The equation shows that 1 mol of aluminium produces $2/4 = 0.5$ mol of aluminium oxide.

Therefore, 0.30 mol produces $0.5 \times 0.30$ mol = 0.15 mol of aluminium oxide.

mass of 1 mol of aluminium oxide = $(2 \times 27) + (3 \times 16) = 102$ g

mass of 0.15 mol = $15 \times 102 = 15.3$ g

Therefore, 15.3 g of aluminium oxide is obtained.

### Worked example 3

Calculate how many tonnes of iron would be produced in a blast furnace from 39.9 tonnes of iron ore, $Fe_2O_3$.

$$2Fe_2O_3(s) + 3C(s) \rightarrow 4Fe(s) + 3CO_2(g)$$

**Answer**

The equation shows that 2 mol of iron(III) oxide react to produce 4 mol of iron. Therefore, 1 mol of iron(III) oxide produces 2 mol of iron.

mass of 1 mol iron(III) oxide = $(2 \times 55.8) + (3 \times 16.0) = 159.6$ g

If 39.9 g had reacted, the number of moles would be:

$$\frac{39.9}{159.6} = 0.25\,mol$$

From the equation, 0.25 mol would produce $2 \times 0.25 = 0.50$ mol of iron.

0.50 mol of iron has mass of $0.5 \times 55.8 = 27.9$ g

∴ 39.9 g of iron(III) oxide produces 27.9 g of iron

Therefore, 39.9 tonnes of iron(III) oxide produces 27.9 tonnes of iron.

*e If questions involve quantities in tonnes, it is easier to do the equivalent calculation in grams and then change this into tonnes. Using tonnes instead of grams is common in exam questions.*

## Questions

**1** What mass of copper(II) oxide is obtained by heating 4.94 g of copper carbonate?

$$CuCO_3(s) \rightarrow CuO(s) + CO_2(g)$$

**2** What mass of zinc chloride is obtained by reacting 3.25 g of zinc with chlorine?

$$Zn(s) + Cl_2(g) \rightarrow ZnCl_2(s)$$

**3 a** Carbon reacts with copper oxide to form copper metal and carbon dioxide. Write the equation for this reaction and calculate the mass of carbon required to react completely with 5 g of copper oxide.

    b Carbon reacts in a similar way with iron(III) oxide to form iron metal.

      (i) Write an equation for this reaction and explain what information is given by this equation.

      (ii) Calculate the mass of carbon required to react completely with 5 g of iron(III) oxide.

    c What mass of carbon is required to react with 5 tonnes of iron(III) oxide?

**4 a** Write the equation, including state symbols, for the reaction of magnesium with zinc sulphate solution to form zinc metal and magnesium sulphate solution.

    b What mass of zinc is obtained by reacting 3 g of magnesium with excess zinc sulphate in solution?

    c Why is it necessary to specify that the zinc sulphate in solution is in excess?

**5** Zinc nitrate decomposes when heated to form zinc oxide:

$$2Zn(NO_3)_2(s) \rightarrow 2ZnO(s) + 4NO_2(g) + O_2(g)$$

    a What information is given by this equation?

    b Calculate the mass of zinc nitrate that would have to be heated to produce 3.6 g of zinc oxide.

**6** Calculate how many tonnes of calcium carbonate must be heated to produce 2800 tonnes of calcium oxide. Carbon dioxide is given off in this reaction.

# Volumes of gases

If a substance is a solid, mass is used to measure the amount needed for a reaction. However, if a reactant or product is a gas, it is unlikely that it could be weighed; it would be easier to measure its volume. In the nineteenth century, the chemist Amedeo Avogadro made this important discovery:

Equal volumes of all gases measured at the same temperature and pressure contain the same number of molecules.

This surprising result is known as **Avogadro's law**. It means that, for example, 50 $cm^3$ of oxygen contains exactly the same number of molecules as 50 $cm^3$ of carbon dioxide, despite the fact that the carbon dioxide molecule is larger and weighs more. This is extremely useful when relating the volume of gas to an equation, as shown in the following worked examples.

---

*Worked example 1*

Assuming the measurements are made at the same temperature and pressure, what volume of oxygen reacts exactly with 100 $cm^3$ of hydrogen?

**Answer**

The equation for the reaction is:

$$2H_2(g) + O_2(g) \rightarrow 2H_2O(l)$$

The mole ratio is 2:1:2.

From the balanced equation, two molecules of hydrogen react with one molecule of oxygen; 2 mol of hydrogen react with 1 mol of oxygen.

The number of molecules in $100\,cm^3$ of hydrogen is not known. However, it is known that $100\,cm^3$ of oxygen contains the same number.

The equation indicates that half the number of oxygen molecules as hydrogen molecules is needed. Therefore, $50\,cm^3$ of oxygen reacts with $100\,cm^3$ of hydrogen.

In worked example 1, if the temperature were such that the $H_2O$ was gaseous, then the volume produced would be $100\,cm^3$. Be careful not to assume that you can simply add the volumes of the reactants to obtain the volume of the product. The total *mass* on the left-hand side of the equation must equal the total mass on the right-hand side, but this does not apply to volumes. In this case, $100\,cm^3$ of hydrogen reacts with $50\,cm^3$ of oxygen but only $100\,cm^3$ of steam is produced.

*e* Remember that the densities of gases vary, and volume = mass/density.

### Worked example 2

Assuming the measurements are made at the same temperature and pressure, what volume of oxygen reacts exactly with $30\,cm^3$ of methane and what volume of carbon dioxide is obtained?

**Answer**

The equation for the reaction is:

$$CH_4(g) + 2O_2(g) \rightarrow CO_2(g) + 2H_2O(l)$$

The mole ratio is 1:2:1:2.

The equation shows that one molecule of methane reacts with two molecules of oxygen to produce one molecule of carbon dioxide and two molecules of water; 1 mol of methane reacts with 2 mol of oxygen to produce 1 mol of carbon dioxide and 2 mol of water.

Therefore, $30\,cm^3$ of methane requires $60\,cm^3$ of oxygen to produce $30\,cm^3$ of carbon dioxide.

*e* As water is a liquid, you cannot deduce its volume using this method.

## Molar volume

If a volume of gas contains a fixed number of molecules, then it is useful to know what volume contains the Avogadro number of particles. This volume is known as the **molar volume**. It depends on the temperature and pressure at which the volume is measured, but there are two useful benchmarks:

■ Under standard conditions of 0°C and 101.3 kPa pressure, the molar volume is $22\,400\,cm^3\,mol^{-1}$. Standard conditions are abbreviated to s.t.p.

■ Under normal laboratory conditions, molar volume is taken as $24\,000\,cm^3\,mol^{-1}$. This value is used in calculations for reactions at room temperature and pressure, although it should be appreciated that it does vary according to the exact conditions. Room temperature and pressure is abbreviated to r.t.p.

Molar volumes can be expressed in $dm^3$:

■ At s.t.p., 1 mol of a gas occupies $22.4\,dm^3$ ($22\,400\,cm^3$)

■ At r.t.p., 1 mol of a gas occupies $24.0\,dm^3$ ($24\,000\,cm^3$)

The key relationship for volumes of gases is:

$$\text{amount of gas (mol)} = \frac{\text{volume of gas}}{\text{molar volume at that temperature and pressure}}$$

At r.t.p., $n = \dfrac{V}{24}$ if V is measured in dm³.

> ℮ In an examination, if you are asked to do a calculation, you will be given the appropriate volume to use at a particular temperature and pressure. The volume at r.t.p. is provided on the data sheet you are given.

---

**Worked example 1**
Calculate the volume of 8 g of methane at r.t.p.

**Answer**
mass of 1 mol of methane, $CH_4 = 12 + (4 \times 1) = 16\,g$

amount in mol of methane $= \dfrac{8}{16} = 0.5\,mol$

volume of 1 mol of methane $= 24\,dm^3$ at r.t.p.
so volume of 0.5 mol methane $= 0.5 \times 24 = 12\,dm^3$

At r.t.p., if volume of gas measured in dm³

---

**Worked example 2**
Calculate the mass of 500 cm³ of oxygen at r.t.p.

**Answer**
volume of 1 mol of oxygen at r.t.p $= 24\,000\,cm^3$

mol in 500 cm³ $= \dfrac{500}{24\,000} = 0.0208\,mol$

mass of 1 mol of oxygen, $O_2 = 32\,g$
so mass of 0.0208 mol $= 0.0208 \times 32 = 0.666\,g$

At r.t.p., if volume of gas measured in cm³

Use either of these along with

To calculate the volume of gas produced in a reaction, use the balanced equation, as shown in the following worked examples. The molar volume appropriate to the conditions of the measurement is always given.

---

**Worked example 1**
Calculate the volume of hydrogen produced when excess dilute sulphuric acid is added to 5.00 g of zinc. Assume that under the conditions of the reaction 1 mol of gas has a volume of 24 000 cm³.

**Answer**
The equation is:

$$Zn(s) + H_2SO_4(aq) \rightarrow ZnSO_4(aq) + H_2(g)$$

The mole ratio is 1:1:1:1.

◀ The term 'excess' indicates that there is enough dilute sulphuric acid for the zinc to react completely.

The balanced equation shows that 1 mol of zinc produces 1 mol of hydrogen molecules. (It also shows that 1 mol of sulphuric acid is used and 1 mol of zinc sulphate is formed, but this is not relevant here.)

mass of 1 mol of zinc = 65.4 g

$$\text{amount in moles of zinc used} = \frac{5.00}{65.4} = 0.0765 \text{ mol}$$

It follows that 0.0765 mol of hydrogen is produced.

molar volume = 24 000 cm³

volume of hydrogen produced = 0.0765 × 24 000 = 1835 cm³

---

## Worked example 2

Calculate the volume of hydrogen produced when excess dilute sulphuric acid is added to 5.00 g of aluminium. Under the conditions of the reaction, 1 mol of gas has a volume of 24 000 cm³.

### Answer

The equation for the reaction is:

$$2Al(s) + 3H_2SO_4(aq) \rightarrow Al_2(SO_4)_3(aq) + 3H_2(g)$$

The mole ratio is 2:3:1:3 or 1:1.5:0.5:1.5.

The equation indicates that 1 mol of aluminium produces 1.5 mol of hydrogen molecules.

$$\text{amount in moles of Al used} = \frac{5.00}{27.0} = 0.185 \text{ mol}$$

amount in moles of $H_2$ produced = 1.5 × 0.185 = 0.278 mol

volume of $H_2$ produced = 0.278 × 24 000 = 6667 cm³

---

## Worked example 3

When reacted with excess dilute hydrochloric acid, what mass of calcium carbonate produces 1 dm³ of carbon dioxide? Assume that 1 mol of gas has a volume of 24 000 cm³.

### Answer

The equation for the reaction is:

$$CaCO_3(s) + 2HCl(aq) \rightarrow 2CaCl_2(aq) + CO_2(g) + H_2O(l)$$

The mole ratio is 1:2:2:1:1.

The equation indicates that 1 mol of carbon dioxide is produced by the reaction of 1 mol of calcium carbonate with excess hydrochloric acid.

$$\text{amount in moles of CO}_2 \text{ produced} = \frac{100}{24\,000} = 4.17 \times 10^{-3} \text{ mol}$$

∴ amount in moles of $CaCO_3$ used = $4.17 \times 10^{-3}$ mol

mass of $CaCO_3$ used = $(4.17 \times 10^{-3})$ × molar mass of $CaCO_3$

mass of 1 mol of $CaCO_3$ = 40.1 + 12.0 + (3 × 16.0) = 100.1 g

mass of $CaCO_3$ used = $(4.17 \times 10^{-3})$ × 100.1 = 0.417 g

## Worked example 4

A gas syringe has a maximum capacity of $100\,cm^3$. Calculate the mass of copper(II) carbonate that would have to be heated to produce enough carbon dioxide to just fill the syringe at r.t.p.

### Answer

The equation for the reaction is:

$$CuCO_3(s) \rightarrow CuO(s) + CO_2(g)$$

The mole ratio is 1:1:1.

syringe volume of $100\,cm^3$ is equivalent to $\dfrac{100}{24\,000} = 0.00417\,mol$

$0.00417\,mol$ of carbon dioxide is obtained by heating $0.00417\,mol$ of copper carbonate

mass of $1\,mol$ of copper carbonate $= 63.5 + 12 + (3 \times 16) = 123.5\,g$

mass of $0.00417\,mol$ of copper carbonate $= 0.00417 \times 123.5 = 0.515\,g$

The answer to worked example 4 is the theoretical mass that could be used. In practice, it would be unwise to heat a mass as large as this because, although the gas when cooled would fit inside the syringe, the hot gas produced would have a volume greater than $100\,cm^3$.

## Questions

**1** Carbon monoxide reacts with oxygen to form carbon dioxide:

$$2CO(g) + O_2(g) \rightarrow 2CO_2(g)$$

What volume of oxygen is required to react with $40\,cm^3$ of carbon monoxide and what volume of carbon dioxide would be formed? Assume that all volumes are measured at the same temperature and pressure.

**2** On heating, dinitrogen monoxide decomposes to nitrogen and oxygen:

$$2N_2O(g) \rightarrow 2N_2(g) + O_2(g)$$

What volumes of oxygen and nitrogen are obtained when $50\,cm^3$ of dinitrogen monoxide decomposes? Assume that all volumes are measured at the same temperature and pressure.

**3** What volume of oxygen reacts exactly with $30\,cm^3$ of propane ($C_3H_8$) and what volume of carbon dioxide is obtained? Assume that all volumes are measured at the same temperature and pressure.

**4** Calculate the volume in $dm^3$ at r.t.p. of each of the following gases:
 **a** $1\,g$ of hydrogen
 **b** $2\,g$ of methane, $CH_4$
 **c** $8.8\,g$ of carbon dioxide
 **d** $4.0\,g$ of nitrogen monoxide, NO

**5** Calculate the mass of each of the following volumes. Assume that the conditions are r.t.p., i.e. $1\,mol$ of gas has a volume of $24\,dm^3$.
 **a** $12\,dm^3$ of oxygen
 **b** $4\,dm^3$ of carbon dioxide
 **c** $500\,cm^3$ of ethane, $C_2H_6$
 **d** $100\,cm^3$ of nitrogen

**6** Calculate the volume of hydrogen in $cm^3$ at r.t.p. that is produced by reacting 1 g of magnesium with excess hydrochloric acid.

**7** Calculate the volume of carbon dioxide in $cm^3$ at r.t.p. that could be obtained when 3 g of carbon is burned in excess oxygen.

**8** Calculate the volume of hydrogen in $cm^3$ at r.t.p. that is needed to convert 7.95 g of copper oxide into copper metal. The equation for the reaction is:

$$CuO(s) + H_2(g) \rightarrow Cu(s) + H_2O(l)$$

**9** Calculate the mass of magnesium that could be reacted with excess sulphuric acid at r.t.p. to fill a 250 $cm^3$ measuring cylinder with hydrogen.

**10** In parts **a–c**, correct any errors that you find.

**a** Consider the reaction:

$$Zn(s) + 2HCl(aq) \rightarrow ZnCl_2(aq) + H_2(g)$$

**(i)** The equation indicates that 1 mol of zinc reacts with 2 mol of hydrochloric acid to produce 1 mol of zinc chloride and 2 mol of hydrogen atoms.

**(ii)** The equation indicates that 1 mol of zinc produces 48 $dm^3$ of hydrogen at room temperature and pressure.

**b** Consider the reaction:

$$CH_4 + 2O_2 \rightarrow CO_2 + 2H_2O$$

The equation indicates that, at room temperature and pressure, 24 $dm^3$ of methane reacts with 48 $dm^3$ of oxygen to produce 24 $dm^3$ of carbon dioxide and 48 $dm^3$ of water.

**c** Gaseous nitrogen monoxide (NO) reacts with oxygen to form gaseous nitrogen dioxide ($NO_2$), according to the equation:

$$2NO(g) + O_2(g) \rightarrow 2NO_2(g)$$

The equation indicates that 2 mol of nitrogen monoxide react with 1 mol of oxygen. So 48 $dm^3$ of nitrogen monoxide reacts with 24 $dm^3$ of oxygen and 72 $dm^3$ of nitrogen dioxide is obtained.

# Summary

You should now be able to:

- convert mass into an amount in moles
- convert an amount in moles into mass
- using a balanced equation, calculate the masses required or obtained in a reaction
- state Avogadro's law
- using a balanced equation, calculate the volumes of gases required and obtained in a reaction

# Additional reading

The French chemist Gay-Lussac did extensive research into the properties of gases. He found by experiment that gases always react in simple ratios of volumes. For example, when hydrogen and oxygen react, the volume of hydrogen required is always twice the volume of oxygen. In the reaction between hydrogen and chlorine, Gay-Lussac found that they react in equal volumes and produce twice the volume of hydrogen chloride. For example, $100\,cm^3$ of hydrogen requires $100\,cm^3$ of chlorine to form $200\,cm^3$ of hydrogen chloride.

Despite the fact that his results appeared to be reliable (Gay-Lussac was a painstakingly careful experimenter), they were not accepted immediately. The difficulty was that Dalton, whose atomic theory had had such an impact on the development of chemical ideas, felt that they conflicted with his understanding of how atoms combine together. The only way that Gay-Lussac's results might be true was if equal volumes of gases contained the same number of atoms — for example, if $100\,cm^3$ of hydrogen contained exactly the same number of atoms as $100\,cm^3$ of chlorine. On reacting, each hydrogen atom would combine with a chlorine atom and this would continue until all the atoms had combined. However, if this were the case, Dalton believed that only $100\,cm^3$ of hydrogen chloride should have been obtained. His logic was based on his belief that the equation for the reaction was H + Cl → HCl, so that every time an atom of hydrogen combined with a chlorine atom, one molecule of hydrogen chloride would be obtained. Dalton then argued that since the final number of molecules of hydrogen chloride obtained was the same as the initial number of hydrogen or chlorine atoms, the hydrogen chloride should have the same volume, i.e. $100\,cm^3$.

It also seemed unlikely that there was a direct relationship between the volume of a gas and the number of particles it contained. However, it was accepted that it might be possible if, in a gas, the atoms or molecules were so far apart that their size was unimportant.

(It is worth appreciating that this type of situation has repeated itself when knowledge has increased and understanding has lagged behind. It is difficult to give up or modify a theory that appears to be well supported by evidence just because a further piece of information casts doubt.)

Avogadro resolved the issue by suggesting that both hydrogen and chlorine exist not as single atoms but as pairs of atoms combined together. The equation then becomes:

$$H_2(g) + Cl_2(g) \rightarrow 2HCl(g)$$

Now the volumes make sense, because the volume of hydrogen chloride produced would be twice that of either the original hydrogen or chlorine.

This was the main reason why scientists discovered that so many gases exist as pairs of atoms joined together in diatomic molecules.

The inclusion of temperature and pressure in Avogadro's law is important because gases readily expand and contract if either of these conditions is altered. The law makes sense only if everything is measured under the same conditions.

Dalton's influence on the scientists of the time was considerable and this hindered the acceptance of the ideas put forward by Avogadro. In fact, Avogadro died before his ideas were accepted as correct.

---

### Joseph Louis Gay-Lussac, 1778–1850

Gay-Lussac was a French scientist of considerable energy. In 1804, he established a world record (held for 50 years) for the highest ascent in a balloon — 21000 feet. By 1809, he was both Professor of Physics at the Sorbonne in Paris and Professor of Chemistry at the École Polytechnique. He established laws concerning combining volumes of gases, established a method for the manufacture of sulphuric acid and, a few days before Davy, discovered boron. (Davy was, however, smart enough to publish his results first!) Gay-Lussac also made wide-ranging contributions to acid–base theory and organic chemistry. Government appointments included a post as superintendent of a gunpowder factory and chief assayer to the mint. In 1831, he was elected to the French House of Representatives.

TOPFOTO

## Extension question: the formula of a gas

The work of Avogadro and Gay-Lussac was important because it provided the means of establishing the formulae of some gases. This question shows you how this might have been done. You have to establish the formula of a gas, $C_xH_y$, from experimental data.

Compound $C_xH_y$ burns in oxygen to form carbon dioxide and water, provided that enough oxygen is supplied.

In an experiment, $40\,cm^3$ of $C_xH_y$ and $500\,cm^3$ of oxygen are mixed together at room temperature and pressure. An electrical spark is then applied, which causes the $C_xH_y$ to react completely. When cooled to room temperature and pressure, droplets of water are observed and the volume of gas is $420\,cm^3$. This gas consists of the carbon dioxide formed and unreacted oxygen.

The $420\,cm^3$ of gas is shaken with sodium hydroxide solution. This absorbs the carbon dioxide by forming sodium carbonate and water. The volume of gas remaining is $260\,cm^3$.

a Write a balanced equation for the reaction between carbon dioxide and sodium hydroxide solution. Include state symbols.

b What volume of carbon dioxide is formed in the reaction?

c What volume of oxygen is used in reacting with $C_xH_y$?

d The equation for the reaction can be written in terms of $x$ and $y$.

$$C_xH_y + \text{......}O_2 \rightarrow x\,CO_2 + \frac{y}{2}H_2O$$

Complete the equation by providing the balancing number for the oxygen molecules.

e Use the volumes you have established in parts b and c and the equation in part d to determine the value of $x$ and then the value of $y$.

f Write the formula of $C_xH_y$.

g Use the balanced equation to determine the mass of water produced when $40\,cm^3$ (measured at r.t.p.) of $C_xH_y$ is burned.

h The density of water is $1\,g\,cm^{-3}$. What volume of water is obtained in the experiment?

# Solutions

## Chapter 5

You should now know how to calculate masses of substances and volumes of gases from balanced equations. However, in experiments, solutions of substances are often used. In this case, it is not possible to find the amount of substance by weighing the solution because this would also include the mass of the solvent (usually water). This chapter explains how the volume of solution can be used instead if the concentration of the solution is known.

A common experiment involves measuring the volume of a solution that reacts with a known volume of another solution. This is called a **titration** or **volumetric analysis**. An accurate titration gives a reliable value for the concentration of a solution.

## Concentrations of solutions

Many chemical reactions involve solutions of substances. The common acids are, for example, usually supplied dissolved in, or diluted with, water. In order to decide on the correct amount to use for a chemical reaction, you need to remember that the weight of a solution includes the mass of the water. Solutions for quantitative work are specified by a concentration that indicates how many grams or, more usefully, moles of the substance there are in a certain volume of solution. This volume is usually $1\,dm^3$ ($1000\,cm^3$).

For example, consider a solution of sodium carbonate containing 2 mol of sodium carbonate in every $1\,dm^3$ of the solution. The concentration is 2 mol per $dm^3$, which is written as $2\,mol\,dm^{-3}$. It follows that $500\,cm^3$ ($0.5\,dm^3$) of this solution contains 1 mol. Different volumes of solution contain different numbers of moles.

In general:

$$\text{amount in moles} = \text{concentration of solution} \times \text{volume of solution}$$
$$(n) \qquad\qquad (\text{mol}\,dm^{-3}) \qquad\qquad (dm^3)$$

$$n = cV$$

> **Worked example 1**
>
> What is the amount in moles of sodium hydroxide in $50\,cm^3$ of a solution of concentration $2\,mol\,dm^{-3}$?
>
> **Answer**
>
> $n = cV$
>
> $2 \times \dfrac{50}{1000} = 0.1\,mol$

To make up a solution of a specific concentration in $mol\,dm^{-3}$ it is necessary to work out the number of grams required to provide the correct number of moles (see Chapter 4).

*e* Remember that the volume in a concentration expression is usually in $dm^3$.

# Diluting solutions

Hydrochloric acid is often sold as a solution of concentration $10\,mol\,dm^{-3}$. If a solution of concentration $2\,mol\,dm^{-3}$ is required, then it has to be diluted five times. For example, $200\,cm^3$ of the concentrated acid requires $800\,cm^3$ water to make $1000\,cm^3$ of dilute acid of concentration $2\,mol\,dm^{-3}$.

A common mistake is to think that $200\,cm^3$ of concentrated acid would require $1000\,cm^3$ of water in order to dilute it five times. That would create $1200\,cm^3$ of solution, which is a six-fold dilution.

## Dilution factor

The dilution factor is calculated using the following formula:

$$\text{dilution factor} = \frac{\text{concentration of the original solution}}{\text{concentration of the diluted solution}}$$

For example, to make $200\,cm^3$ of a solution of concentration $0.5\,mol\,dm^{-3}$ from a stock solution of concentration $2\,mol\,dm^{-3}$:

$$\text{dilution factor} = \frac{2}{0.5} = 4$$

Therefore, the stock solution has to be diluted four times. To make $200\,cm^3$ of concentration $0.5\,mol\,dm^{-3}$ requires $50\,cm^3$ of the stock solution and $150\,cm^3$ of water.

Diluting concentrated acids is an exothermic process. If a small volume of water is added to concentrated acid, the heat produced may cause the water to boil, which can be dangerous. Concentrated acid should always be added slowly to water; water should *never* be added to concentrated acid.

e Remember *AAA* — **a**lways **a**dd **a**cid.

---

### Worked example 1

How can $500\,cm^3$ of a $0.4\,mol\,dm^{-3}$ solution of sulphuric acid be prepared from a solution of concentration $2\,mol\,dm^{-3}$?

**Answer**

$$\text{dilution factor} = \frac{2}{0.4} = 5$$

Therefore, $100\,cm^3$ of the dilute sulphuric acid of concentration $2\,mol\,dm^{-3}$ should be further diluted by the addition of $400\,cm^3$ of water.

---

### Worked example 2

What is the concentration in $mol\,dm^{-3}$ of $375\,cm^3$ of dilute sodium hydroxide solution prepared by diluting $75\,cm^3$ of sodium hydroxide solution of concentration $0.6\,mol\,dm^{-3}$ with water?

**Answer**

$$\text{volume of dilute solution} = 375\,cm^3$$

$$\text{concentration} = \frac{75}{375} \times 0.6 = 0.12\,mol\,dm^{-3}$$

---

## Questions

**1** State the amount in moles in each of the following:
 **a** $2\,dm^3$ of sodium hydroxide solution of concentration $0.1\,mol\,dm^{-3}$
 **b** $200\,cm^3$ of sulphuric acid of concentration $0.5\,mol\,dm^{-3}$
 **c** $50\,cm^3$ of sodium carbonate solution of concentration $0.05\,mol\,dm^{-3}$
 **d** $25\,cm^3$ of hydrochloric acid of concentration $0.1\,mol\,dm^{-3}$

**2** Calculate the volume of $0.5\,mol\,dm^{-3}$ potassium chloride solution that contains:
 **a** $1\,mol$
 **b** $0.2\,mol$
 **c** $0.05\,mol$
 **d** $0.36\,mol$

**3** Give the concentration of each of the following in $mol\,dm^{-3}$:
 **a** $100\,cm^3$ containing $0.1\,mol$
 **b** $50\,cm^3$ containing $0.01\,mol$
 **c** $5\,dm^3$ containing $2\,mol$
 **d** $25\,cm^3$ containing $0.027\,mol$

**4** Calculate the mass of solid that on dissolving would produce:
 **a** $250\,cm^3$ of sodium hydroxide solution of concentration $0.2\,mol\,dm^{-3}$
 **b** $100\,cm^3$ of sodium carbonate solution of concentration $0.4\,mol\,dm^{-3}$
 **c** $50\,cm^3$ of potassium sulphate solution of concentration $0.15\,mol\,dm^{-3}$
 **d** $75\,cm^3$ of sodium chloride solution of concentration $0.4\,mol\,dm^{-3}$

**5** Describe how each of the following solutions of sulphuric acid could be prepared from a solution of acid of concentration $1 \, mol \, dm^{-3}$:

  **a** $1 \, dm^3$ of solution of concentration $0.1 \, mol \, dm^{-3}$

  **b** $500 \, cm^3$ of solution of concentration $0.2 \, mol \, dm^{-3}$

  **c** $100 \, cm^3$ of solution of concentration $0.15 \, mol \, dm^{-3}$

**6** What is the concentration obtained by diluting $50 \, cm^3$ of $2 \, mol \, dm^{-3}$ sodium carbonate solution with water to make:

  **a** $500 \, cm^3$ of solution

  **b** $650 \, cm^3$ of solution

  **c** $320 \, cm^3$ of solution

# Titrations

In a titration a solution is added gradually to a known volume of another solution until the exact volume required to complete the reaction is reached. If the concentration of one solution is known then, using the balanced equation for the reaction, the concentration of the other can be deduced. An example is the titration of an alkali with an acid, which requires the use of an acid–base indicator, such as methyl orange or phenolphthalein, to show when the reaction is complete.

Using a pipette and a burette to deliver precise quantities of solutions allows a titration to be carried out accurately.

*Performing an acid–base titration*

MARTYN F. CHILLMAID/SPL

For example, using a pipette, exactly $25.0 \, cm^3$ of sodium hydroxide solution of concentration of $0.100 \, mol \, dm^{-3}$ is taken. This is transferred to a conical flask and a few drops of methyl orange indicator are added. Sodium hydroxide solution is alkaline, so the methyl orange appears yellow. Hydrochloric acid of unknown concentration is added carefully from a burette until the indicator turns orange. This shows that the neutral point has been reached. The volume of hydrochloric acid added is noted. The titration is repeated until consistent results are obtained. With care, it is possible to obtain reproducible results that are accurate to within $0.1 \, cm^3$.

The point at which the two solutions have reacted completely is called the **end point** of the titration; the volume of solution added from the burette is known as the **titre**.

## Tabulating the results of a titration

The results of a titration should be recorded in an orderly way. It is best to tabulate the data. For each titration, record the initial burette reading and the final reading, and then calculate the volume used.

Table 5.1 shows how the the results of a titration should be recorded for the above example.

| Titration | Rough | 1 | 2 | 3 |
|---|---|---|---|---|
| Final burette reading (cm³) | 19.10 | 18.55 | 18.85 | 37.50 |
| Initial burette reading (cm³) | 0.00 | 0.00 | 0.00 | 18.85 |
| Volume used (cm³) | 19.10 | 18.55 | 18.85 | 18.65 |

*Table 5.1 Results for a titration*

A burette is usually accurate to $0.05 \, cm^3$ and the results recorded should indicate this. It is usual, but not essential, to do a 'rough' titration first as a guide to the end point of the titration. This is then followed by more accurate readings, obtained by adding the solution from the burette dropwise as the end point is approached. Some students record the rough titration to one decimal place only, since it is done with less precision than the other titrations. The 'rough' value is not used in working out the titre.

The results in Table 5.1 show that the titres from the first and third accurate titrations are in close agreement. The titre from the second accurate titration is out of line and is not used in calculating the mean titre. Therefore, the average value for the results of this titration is $18.60 \, cm^3$.

## Calculation of the unknown concentration

The concentration of the hydrochloric acid in the above example can be calculated as follows.

**Step 1:** calculate the amount (in moles) of sodium hydroxide used.

This is possible because the concentration, $c$, and the volume, $V$, are known.

**ⓔ** The correct use of pipettes and burettes and the limits of their accuracy are covered in Chapter 14.

**ⓔ** Do not forget to write the initial reading, even if it is zero, to the same level of accuracy as the final reading.

**ⓔ** The solutions in the conical flask should be mixed together carefully.

**ⓔ** Calculations of this type appear frequently in exams and may also be part of your school practical assessment.

concentration of sodium hydroxide = $0.100\,mol\,dm^{-3}$ (per $1000\,cm^3$)

$25.0\,cm^3$ was taken in the pipette

$25.0\,cm^3$ of sodium hydroxide contains $0.100 \times \dfrac{25.0}{1000} = 0.00250\,mol$

$(n = cV = 0.100 \times 0.0250)$

**Step 2:** refer to the balanced equation to find the amount in moles of hydrochloric acid needed to react with $0.00250\,mol$ of sodium hydroxide.

The equation for the reaction is:

$$NaOH + HCl \rightarrow NaCl + H_2O$$

The mole ratio is 1:1:1:1.

Therefore, 1 mol of NaOH reacts with 1 mol of HCl.

So $0.00250\,mol$ of HCl are needed to react completely with $0.00250\,mol$ NaOH.

Since $18.60\,cm^3$ of HCl was added from the burette, it follows that $18.60\,cm^3$ contains $0.00250\,mol$ of acid.

**Step 3:** calculate the concentration of hydrochloric acid in $mol\,dm^{-3}$.

volume of hydrochloric acid = $18.60\,cm^3 = \dfrac{18.60}{1000}\,dm^3$

amount (in moles) of hydrochloric acid = $0.00250\,mol$

concentration of hydrochloric acid = $\dfrac{1000}{18.60} \times 0.00250\,mol\,dm^{-3}$

$$= 0.134\,mol\,dm^{-3}$$

---

*Worked example 1*

$25.00\,cm^3$ of sodium carbonate is neutralised by $27.20\,cm^3$ of hydrochloric acid of concentration $0.120\,mol\,dm^{-3}$. What is the concentration in $mol\,dm^{-3}$ of the sodium carbonate solution?

**Answer**

The concentration, $c$, and the volume, $V$, of the hydrochloric acid are known.

concentration of hydrochloric acid = $0.120\,mol\,dm^{-3}$

$27.20\,cm^3$ of acid contains $\dfrac{27.20}{1000} \times 0.120\,mol = 0.003264\,mol$

The equation for the reaction is:

$$Na_2CO_3 + 2HCl \rightarrow 2NaCl + H_2O + CO_2$$

The mole ratio is 1:2:2:1:1.

1 mol of sodium carbonate reacts with 2 mol of hydrochloric acid, so for every 1 mol of hydrochloric acid that reacts, 0.5 mol of sodium carbonate is required. $0.003264\,mol$ of hydrochloric acid reacts with $0.5 \times 0.003264 = 0.001632\,mol$ of sodium carbonate.

This amount must have been present in the $25.00\,cm^3$ of sodium carbonate used.

concentration of sodium carbonate = $\dfrac{1000}{25.0} \times 0.001632 = 0.0653\,mol\,dm^{-3}$

*e* You need to be able to do such calculations without help. However, exam questions will give you some guidance.

It is sometimes possible to deduce further information from a titration, as illustrated by the following worked example.

### Worked example

5.00 g of an impure sample of potassium carbonate is weighed out and dissolved to make 1.00 dm$^3$ of solution. 25.0 cm$^3$ of this solution is titrated against 0.050 mol dm$^{-3}$ hydrochloric acid. 28.80 cm$^3$ of the hydrochloric acid is needed to neutralise the potassium carbonate solution.

**a** Use the titration results to calculate the concentration of the potassium carbonate solution.

**b** Deduce the percentage purity of the solid sample of potassium carbonate.

### Answer

**a** concentration of hydrochloric acid = 0.050 mol dm$^{-3}$

28.80 cm$^3$ of hydrochloric acid contains

$$\frac{28.80}{1000} \times 0.050 \, \text{mol} = 0.00144 \, \text{mol}$$

The equation for the reaction is:

$$K_2CO_3 + 2HCl \rightarrow 2KCl + H_2O + CO_2$$

The mole ratio is 1:2:2:1:1.

1 mol of potassium carbonate reacts with 2 mol of hydrochloric acid, so for every 1 mol of hydrochloric acid that reacts, 0.5 mol of potassium carbonate is required. 0.00144 mol of hydrochloric acid reacts with $0.5 \times 0.00144 = 0.000720$ mol of potassium carbonate.

This amount must have been present in the 25.00 cm$^3$ of potassium carbonate used.

$$\text{concentration of potassium carbonate} = \frac{1000}{25.00} \times 0.000720$$

$$= 0.0288 \, \text{mol dm}^{-3}$$

**b** relative formula mass of potassium carbonate, K$_2$CO$_3$

$$= (2 \times 39.1) + 12.0 + (3 \times 16.0) = 138.2$$

mass of pure K$_2$CO$_3$ in the 0.0288 mol dm$^{-3}$ solution

$$= 0.0288 \times 138.2 = 3.98 \, \text{g}$$

Therefore, this must be the mass of pure potassium carbonate in the original 5.00 g.

$$\text{percentage purity} = \frac{3.98}{5.00} \times 100 = 79.6\%$$

## Questions

**1** A solution of potassium hydroxide has a concentration of 0.100 mol dm$^{-3}$. 25.0 cm$^3$ of this solution is neutralised by 22.50 cm$^3$ of hydrochloric acid.

**a** Write a balanced equation for the reaction between potassium hydroxide and hydrochloric acid.

**b** What amount in moles of potassium hydroxide is contained in 25.0 cm$^3$ of solution?

**c** What amount in moles of hydrochloric acid will neutralise the potassium hydroxide present in 25.0 cm³ of solution?

**d** What is the concentration of the hydrochloric acid in mol dm⁻³?

2 A solution of potassium carbonate has a concentration of 0.04 mol dm⁻³. 25.0 cm³ of this solution is neutralised by 28.10 cm³ of hydrochloric acid.

    **a** Write a balanced equation for the reaction between potassium carbonate and hydrochloric acid.

    **b** What amount in moles of potassium carbonate is contained in 25.0 cm³ of solution?

    **c** What amount in moles of hydrochloric acid is needed to neutralise the number of moles of potassium carbonate in 25.0 cm³ of solution?

    **d** What is the concentration of the hydrochloric acid in mol dm⁻³?

3 Magnesium oxide neutralises nitric acid. In an experiment, it is found that 1.20 g of magnesium oxide is neutralised by 24.20 cm³ of nitric acid.

    **a** Write an equation for the reaction between magnesium oxide and nitric acid.

    **b** What is the amount in moles of 1.20 g of magnesium oxide?

    **c** What amount in moles of nitric acid is contained in 24.20 cm³?

    **d** What is the concentration of the nitric acid in mol dm⁻³?

4 **a** Write an equation for the reaction between zinc oxide and hydrochloric acid.

    **b** What is the amount in moles of 4.07 g of zinc oxide?

    **c** Calculate the volume of 2 mol dm⁻³ hydrochloric acid that would react exactly with 4.07 g of zinc oxide.

    **d** If 2 mol dm⁻³ sulphuric acid is used instead of 2 mol dm⁻³ hydrochloric acid, what volume of sulphuric acid would be required?

5 100 cm³ of hydrochloric acid is diluted to form 1 dm³ of solution. When the diluted acid is titrated against 0.500 mol dm⁻³ sodium carbonate solution, it is found that 27.20 cm³ of the diluted hydrochloric acid is needed to neutralise 25.0 cm³ of the sodium carbonate solution.

    **a** Write the equation for the reaction.

    **b** Calculate the amount in moles of sodium carbonate in 25.0 cm³ of solution.

    **c** Calculate the concentration of the diluted hydrochloric acid.

    **d** What is the concentration of the hydrochloric acid before dilution?

6 A solution of potassium hydroxide contains 5.60 g dm⁻³. This solution is used to neutralise 20.0 cm³ of a solution of sulphuric acid of concentration 0.05 mol dm⁻³.

    **a** Write an equation for the reaction.

    **b** Determine the volume of potassium hydroxide solution that is required to neutralise the sulphuric acid.

7 A solution is made by dissolving 5.00 g of sodium hydroxide to make 1 dm³ of solution. A solution of an acid, $H_2X$, is made by dissolving 7.66 g to make 1 dm³ of solution. 25.0 cm³ of the solution of sodium hydroxide is neutralised by 20.0 cm³ of the solution of the acid $H_2X$.

    **a** Write an equation for the reaction.

    **b** Calculate the concentration of the sodium hydroxide in mol dm⁻³.

    **c** Calculate the amount in moles of sodium hydroxide in 25.0 cm³.

    **d** Calculate the amount in moles of $H_2X$ in 20.0 cm³ of solution.

    **e** Calculate the concentration of $H_2X$ in mol dm⁻³.

    **f** Calculate the mass of the anion, $X^{2-}$.

**8** 10.00g of sodium carbonate crystals $Na_2CO_3 \cdot xH_2O$ are dissolved to make 1 $dm^3$ of solution. 25.0 $cm^3$ of this solution is neutralised by 17.50 $cm^3$ of hydrochloric acid of concentration 0.100 $mol\,dm^{-3}$.

    **a** Calculate the amount in moles of hydrochloric acid present in 17.50 $cm^3$.

    **b** What is the amount in moles of sodium carbonate present in 25.0 $cm^3$ of solution?

    **c** Calculate the concentration in $mol\,dm^{-3}$ of the sodium carbonate solution.

    **d** Use your answer to **c** to calculate the molar mass of sodium carbonate crystals.

    **e** Deduce the value of $x$.

# Summary

You should now be able to:

- understand what is meant by the concentration of a solution
- convert concentrations in $g\,dm^{-3}$ to $mol\,dm^{-3}$ and vice versa
- relate concentrations in $mol\,dm^{-3}$ and volumes to the number of moles present
- calculate the dilution necessary to convert one concentration into another
- use the results of a titration to calculate the concentration of a solution

# Additional reading

An indicator is used in an acid–base titration to show when the end point of a reaction between two colourless solutions is reached. Indicators exploit a common property of coloured organic compounds that contain large numbers of double bonds and, sometimes, pairs of electrons that are not involved in bonding. These molecules can bond with hydrogen ions to create a new species with a different colour. Some pigments that give colour to plants do this and it is possible to make an adequate indicator by extracting such a colour. When the compound is in alkaline solution there are no hydrogen ions to absorb, but as the alkali is neutralised and excess acid added, the indicator combines with the hydrogen ions and there is a colour change.

There are two important considerations in choosing an indicator. First, the colour change must be easy to see. In practice, it may be difficult to distinguish between closely related colours, such as a change from blue to purple (the colour change of litmus) or yellow to orange (the colour change of methyl orange). Colour blindness may make other changes hard to identify.

The second consideration relates to the exact level of acidity needed for the indicator to respond. It is rare for an indicator to change exactly at the true neutral point of pH 7. For example, methyl orange changes only when slightly acidic (pH range 3.0–4.5); phenolphthalein responds when the solution is slightly alkaline (pH range 8.0–10.0). A suitable indicator has to be chosen for each titration.

Titrations are not restricted to reactions between acids and alkalis. Any reaction with a colour change can be used for volumetric analysis. A few reagents, such as potassium manganate(VII), are self-indicating. In some reactions, the intense purple colour of potassium manganate(VII) changes to pale pink as manganate(VII) is reduced to manganese(II). For other reactions, indicators may be required. For example, the colour change from orange potassium dichromate(VI) to green chromium(III) as dichromate(VI) is not easy to see but can be enhanced by an indicator so that titration becomes possible.

Another useful technique is a back titration. Suppose we want to determine the atomic mass of an element, X, that reacts with sulphuric acid according to the equation:

$$X + H_2SO_4 \rightarrow X_2SO_4 + H_2$$

A known mass of X is reacted with excess sulphuric acid of known volume and concentration. After the reaction, a titration is used to establish how much sulphuric acid is left unreacted.

From the initial volume and concentration of the sulphuric acid and the result of the titration, it is possible to establish how many moles of sulphuric acid had reacted. From the equation it can be seen that this is the same as the number of moles of X. A known mass of X was used and the amount (in moles) of X is now known. Therefore, it is possible to deduce the atomic mass of X.

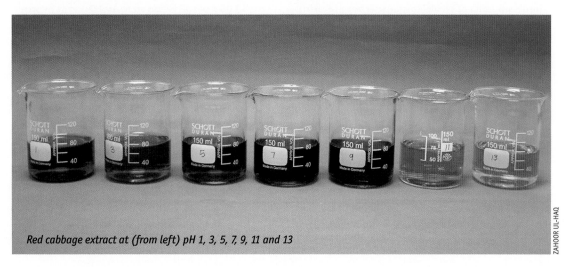

*Red cabbage extract at (from left) pH 1, 3, 5, 7, 9, 11 and 13*

ZAHOOR UL-HAQ

*On the beach near Beachy Head*

ANDREW HOLT/ALAMY

Beachy Head, near Eastbourne in East Sussex, is a dominating chalk cliff characteristic of the scenery of that part of the south coast. The soft chalk is subject to erosion by the sea. Areas of the cliff edge have to be cordoned off when they become unstable; they will eventually fall into the sea. Constant monitoring of the state of the cliff is essential to ensure public safety.

A soil sample taken from the area consists of large amounts of chalk (calcium carbonate) with some clay and other silicate materials, as well as decomposing organic matter. Assuming that the chalk is the only material that reacts with dilute acid, there are several ways of determining the percentage of chalk in the soil sample.

1 *Collection of carbon dioxide*

A 1.00 g sample of a soil is reacted with excess hydrochloric acid and the carbon dioxide evolved is collected over water. After cooling to room temperature, 89 cm$^3$ of carbon dioxide is obtained.

a Write a balanced equation for the reaction between calcium carbonate and hydrochloric acid.

b Calculate the amount in moles of carbon dioxide obtained.

c Deduce the number of moles of calcium carbonate that react.

d Calculate the mass of calcium carbonate that reacts.

e Calculate the percentage of chalk in the soil sample.

2 *Back titration*

10.00 g of soil is added to 100.0 cm$^3$ of hydrochloric acid of concentration 1.00 mol dm$^{-3}$. Once all the chalk has reacted, the residue is allowed to settle. Excess acid remains in the solution. Using a pipette, 25.00 cm$^3$ of this solution is carefully removed and titrated against 0.0500 mol dm$^{-3}$ sodium carbonate solution. 23.10 cm$^3$ is required to neutralise the acid.

a Write a balanced equation for the reaction between sodium carbonate and hydrochloric acid.

b Calculate the amount in moles of sodium carbonate in the 23.10 cm$^3$ required by the titration.

c Use your answer to b to deduce the amount in moles of hydrochloric acid contained in 25.00 cm³ of the solution.

d Deduce the amount in moles in 100.0 cm³ of the hydrochloric acid after it reacts with the soil sample.

e Calculate the amount in moles of hydrochloric acid in the 100.0 cm³ before it reacts with the soil sample.

f Use your answers to d and e to deduce the amount in moles of hydrochloric acid that react with the soil sample.

g Write the equation for the reaction between calcium carbonate and hydrochloric acid.

h Use the equation in g and your answer to f to state the amount in moles of calcium carbonate in 10.00 g of soil.

i Deduce the mass of calcium carbonate in 10.00 g of soil.

j Calculate the percentage of chalk in the soil sample.

3 Consider the experimental methods in Questions 1 and 2. Suggest reasons for the difference in the answers for the percentage of chalk in the soil sample obtained by the two methods.

4 Two other possible methods of determining the percentage of chalk in a sample of the soil are:

■ A weighed sample of the soil is heated strongly and the volume of carbon dioxide evolved is measured.

■ A weighed sample of the soil is reacted with dilute hydrochloric acid until all the chalk dissolves. The solution is filtered and the mass of the unreacted residue is found.

Consider each of these methods and suggest, with reasons, whether they would be likely to produce an accurate result.

# Bonding and structure

## Chapter 6

Atoms can combine in two main ways:

■ by losing or gaining electrons (**ionic bonding**)
■ by sharing electrons (**covalent bonding**)

However, there is a sliding scale from substantially ionic character to total covalency.

In addition, molecules themselves have some capacity to attract each other, which is why it is possible to cause substances to liquefy or become solid. For some substances, the interaction between the molecules is weak and extremely low temperatures are required to solidify them. For others, the interaction can be strong enough to affect their physical properties. The solids formed also differ. They can be large, crystalline structures of considerable strength or weakly bonded substances with no specific shape. Understanding the principles of structure and bonding is essential for a full appreciation of both organic and inorganic chemistry.

## Background

Chapter 2 introduced the concept that the mobility of electrons is responsible for the ways in which atoms react to form compounds. Valency was used to provide a simple method of establishing the formulae of compounds. In this chapter, the nature of the bonds holding compounds together is discussed in detail. As chemical bonding involves the interaction of outer-shell electrons (i.e. those of the highest principal quantum number) it is important that you review the aspects of electron configuration covered in Chapter 1.

There are three ways in which atoms bond together chemically:

■ ionic bonding
■ covalent bonding
■ metallic bonding

All three types of bond are strong and a large amount of energy is required to break them.

---

### Question

Using the *s*, *p*, *d* notation, write the full electron configuration of the following atoms or ions:

**a** sodium atom, Na
**b** sodium ion, $Na^+$
**c** oxygen atom, O

**d** oxide ion, $O^{2-}$
**e** aluminium atom, Al
**f** aluminium ion, $Al^{3+}$

---

You should note that, except for helium, the noble gases have the same outer-shell electron configuration ($ns^2\, np^6$). The full electron configurations of the noble gases neon, argon and krypton are:

- $_{10}$Ne — $1s^2\, 2s^2\, 2p^6$
- $_{18}$Ar — $1s^2\, 2s^2\, 2p^6\, 3s^2\, 3p^6$
- $_{36}$Kr — $1s^2\, 2s^2\, 2p^6\, 3s^2\, 3p^6\, 3d^{10}\, 4s^2\, 4p^6$

An atom of a noble gas contains eight electrons in its outer shell (two $s$-electrons and six $p$-electrons). The noble gases are not reactive; in the early twentieth century it was suggested that the lack of reactivity is associated with electron configuration. This led to the suggestion that when atoms of elements combine to form compounds, they do so by gaining, losing or sharing electrons to attain the same number of electrons as a noble gas. This is known as the **octet rule**. Lighter elements, for example hydrogen and lithium, produce compounds with two electrons in their outer shells, so they have the same electron configuration as helium. For many compounds the octet rule is broken because some elements have more than eight electrons in their outer shells. Nevertheless, the octet rule is helpful in establishing how elements combine.

Another way of stating the octet rule is to say that when atoms bond together, the elements in the resultant compounds tend to be **isoelectronic** with a noble gas.

◀ Isoelectronic means 'the same number of electrons'.

The capacity of an atom to form chemical bonds is the valency of the element. This depends on the number of electrons in the outer electron (valence) shell. The periodic table can be used as a simple guide to predict the valency and, therefore, the number of chemical bonds formed by an element (Table 6.1).

**Table 6.1 The periodic table and valency**

|  | Group 1 | Group 2 | Group 3 | Group 4 | Group 5 | Group 6 | Group 7 |
|---|---|---|---|---|---|---|---|
| Period 3 | Na | Mg | Al | Si | P | S | Cl |
| Valency (number of chemical bonds) | 1 | 2 | 3 | 4 | 3 | 2 | 1 |

There are exceptions, so this table should be used only as a rough guide.

# Ionic bonding

An ionic bond is formed by the electrostatic attraction between oppositely charged ions.

Ionic bonds are usually formed between a metal and a non-metal. A metal atom loses one or more electrons from its outer (valence) shell and the electrons are transferred to the non-metal atom or atoms. The resultant ions each have the electronic structure of a noble gas. (Apart from helium, this means eight electrons in their outer shells.)

Sodium chloride is an ionic compound formed by the transfer of an electron from each sodium atom to each chlorine atom.

Sodium is element 11 ($_{11}$Na). A sodium atom has the electron configuration $1s^2\ 2s^2\ 2p^6\ 3s^1$, with one electron in its outer third shell. Using the octet rule, you can see that by losing its single outer-shell electron, sodium attains the electron configuration $1s^2\ 2s^2\ 2p^6$. As it has lost an electron, it is now an ion. As an atom it had 11 protons and 11 electrons. One electron has been lost, so the ion has 11 protons but only ten electrons. Therefore, it has a net charge of 1+.

Chlorine is element 17 ($_{17}$Cl). A chlorine atom has the electron configuration $1s^2\ 2s^2\ 2p^6\ 3s^2\ 3p^5$, with seven electrons in its outer third shell. The octet rule predicts that chlorine will gain one electron to achieve the electron configuration $1s^2\ 2s^2\ 2p^6\ 3s^2\ 3p^6$. As it has gained an electron, it is now an ion with 18 electrons and 17 protons. Therefore, it has a charge of 1−.

The formation of sodium chloride can be shown by a **dot-and-cross** diagram. Dot-and-cross diagrams do not give the orbital structures. They are simplified structures showing the electrons in rings around the nucleus. They are useful because they convey the basic information about bonding in a simple, visual way.

According to the octet rule, a sodium atom needs to lose one electron from the outer shell and a chlorine atom needs to gain one electron:

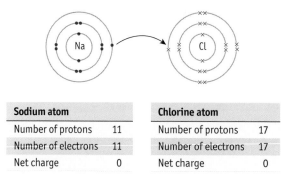

| Sodium atom | |
| --- | --- |
| Number of protons | 11 |
| Number of electrons | 11 |
| Net charge | 0 |

| Chlorine atom | |
| --- | --- |
| Number of protons | 17 |
| Number of electrons | 17 |
| Net charge | 0 |

The resultant sodium and chloride ions each have eight electrons in their outer shells.

| Sodium ion | |
| --- | --- |
| Number of protons | 11 |
| Number of electrons | 10 |
| Net charge | 1+ |

| Chloride ion | |
| --- | --- |
| Number of protons | 17 |
| Number of electrons | 18 |
| Net charge | 1− |

It is important to note that the ions are separate from each other. They are held together solely by the electrical attraction between them.

**e** In exam questions, it is usual to show the outer-shell electrons only.

Ions formed from metal atoms are always created by the loss of electrons and are, therefore, positively charged. They are called **cations**. Ions formed from non-metal atoms by gaining electrons are negatively charged. They are called **anions**. The oppositely charged ions are attracted to each other and this constitutes the ionic bond.

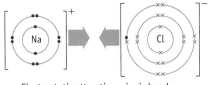

Electrostatic attraction = ionic bond

When sodium and chlorine react together, vast numbers of atoms react simultaneously. Each positive sodium ion produced is not attracted to just one chloride ion. It is attracted to other surrounding chloride ions; each chloride ion is attracted to other surrounding sodium ions. In fact, each sodium ion is attracted to six chloride ions and each chloride ion is attracted to six sodium ions:

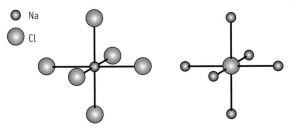

In this way, the ions of sodium chloride build into a crystal structure. In sodium chloride, each ion is bonded ionically to six oppositely charged ions:

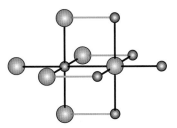

A giant ionic lattice is formed, which is said to be '6:6 coordinate'. In sodium chloride, the shape of the crystal is cubic:

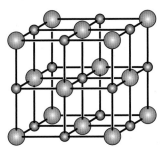

*Figure 6.1 Sodium chloride ionic lattice*

Every ionic compound forms a giant ionic lattice that is held together by strong electrostatic attractions (ionic bonds) throughout the lattice. The ionic bond is directional and determines the shape of the lattice. The ions are held in a fixed position within the lattice and are not free to move.

**e** There are a number of different possible crystal structures.

## Question

Magnesium oxide and calcium fluoride both have ionic bonds. For each compound:
a draw dot-and-cross diagrams to show the atoms and ions
b write the electron configurations of the atoms and ions

## Typical properties of ionic compounds

As a result of their structure, ionic compounds are solids with high melting points. When solid they do not conduct electricity. However, they do conduct electricity when they are molten or dissolved in water. These properties can be explained by the nature of the ionic bond.

An ionic compound has a high melting point because ionic bonds are strong and occur throughout the giant lattice. In order to melt an ionic solid, all the ionic bonds in the lattice have to be broken. This requires a large amount of energy. Hence, the melting point is high.

An ionic compound does not conduct electricity when solid because the ions are held in a fixed position within the lattice and cannot move. To conduct electricity, mobile charge carriers (free electrons or free ions) are required. When an ionic compound is molten or dissolved in water, the giant ionic lattice is broken down and the ions are free to move. Therefore, in an ionic compound, the mobile charge carriers are the free ions — the molten solid conducts electricity. Positive ions move to the cathode — hence the name cation. Negative ions (anions) move to the anode.

You may find it surprising that most ionic compounds dissolve in water, since the strong ionic bonds of the lattice have to be broken. The ability of water to permeate some ionic lattices and break them apart is discussed on page 31, when the nature of the bonds in water is considered. For now, it is important that you appreciate that when ionic compounds are dissolved, the individual ions become separated. Therefore, the solution can conduct electricity. The ions act independently of each other, which is shown by their chemical reactions. For example, aqueous chloride ions from dissolved sodium chloride react with aqueous silver ions from dissolved silver nitrate to produce a precipitate of silver chloride. This is discussed in the section on ionic equations (page 32).

Be careful not to confuse the reactivity of atoms with the reactivity of ions. The sodium atom is reactive because, in chemical reactions, it forms an ion readily. Conversely, the sodium ion is unreactive because it is not easily converted back into a sodium atom.

Metals conduct electricity by the movement of free electrons; ionic compounds conduct electricity only if the ions are free to move.

# Covalent bonding

A covalent bond is formed by the sharing of two electrons between two adjacent atoms. Each atom provides one of each shared pair of electrons.

Covalent bonds are usually formed between two non-metal atoms. Both non-metal atoms share electrons from their outer (valence) shell. Each atom achieves the electron configuration of a noble gas, which, apart from helium, is eight electrons in the outer shell.

Many compounds in everyday life are covalent. Almost all the compounds in food and most of the compounds in our bodies are bonded covalently. Covalent compounds usually exist as simple discrete molecules, but some form giant structures (page 95).

## Single covalent bonds

Some non-metallic elements exist as pairs of atoms linked together. The bonding is covalent and occurs because, by sharing electrons, each atom achieves the stability of a noble gas configuration. A **single covalent bond** is formed from the sharing of two electrons by adjacent atoms.

Each hydrogen atom has one electron in its outer orbit:

A hydrogen molecule is a simple diatomic molecule in which the atoms are held together by a single covalent bond:

The hydrogen molecule can be represented as H–H.

A chlorine molecule is also a diatomic molecule with the atoms held together by a single covalent bond. Each chlorine atom has seven electrons in its outer orbit and is, therefore, one electron short of attaining the octet:

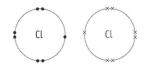

By sharing electrons, each chlorine atom obeys the octet rule:

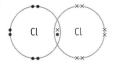

The chlorine molecule can be represented as Cl–Cl.

Covalent bonds also occur between atoms of different non-metallic elements. For example, methane is a simple covalent compound. A molecule of methane consists of one carbon atom and four hydrogen atoms. A carbon atom requires four electrons to be isoelectronic with a noble gas; each hydrogen atom requires one electron:

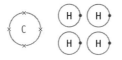

By sharing electrons, each atom in a molecule of methane achieves a stable electron configuration:

The methane molecule can be drawn as follows:

In dot-and-cross diagrams that show covalent bonding, the atoms are joined physically by the sharing of electrons. This is in contrast to ionic compounds, in which the ions are separate but held in place by electrical attraction.

## Multiple covalent bonds

### Double bonds

If two atoms share four electrons, two covalent bonds are formed — a **double bond**. The simplest substance containing a double bond is an oxygen molecule. Each oxygen atom has six electrons in its outer shell and is two electrons short of attaining the octet:

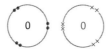

By sharing two of its electrons, each oxygen atom satisfies the octet rule:

The oxygen molecule can be represented as O=O.

Double bonds are often referred to as **unsaturated bonds** and are common in organic molecules. For example, ethene, $C_2H_4$, and methanal, HCHO, are unsaturated compounds. The diagrams below show the single and double covalent bonds in these compounds:

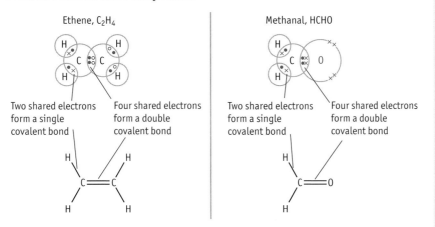

## Triple bonds

Nitrogen molecules contain a **triple** covalent bond, formed by sharing six electrons. Each nitrogen atom has five electrons in its outer shell and is three electrons short of attaining the octet:

By sharing three of its electrons, each nitrogen atom attains a stable outer octet:

The nitrogen molecule can be represented as:

$$N \equiv N$$

# Dative covalent (coordinate) bonding

When adjacent atoms bond together and share two electrons, it is usual for each atom to provide one electron. Sometimes, however, one atom supplies *both* electrons. Two electrons are shared, so the bond is covalent, but, because both shared electrons come from one of the atoms, the bond is called a **dative covalent bond** (coordinate bond).

A simple example of a dative covalent bond also explains how ammonia ($NH_3$) forms the ammonium ion, $NH_4^+$. This is possible because the nitrogen atom in ammonia has eight electrons in its valence shell, two of which are not bonded to other atoms. This pair of electrons is called a **lone pair**.

If a hydrogen ion (which has no electrons) is provided by an acid, the lone pair of electrons on the nitrogen atom can form a dative covalent bond with the hydrogen ion:

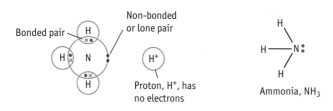

An ammonium ion, which has a charge of 1+, is formed:

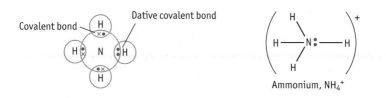

In an ammonium ion, all four bonds from nitrogen to hydrogen are equivalent. It is not possible to identify which is the dative covalent bond. It is simply the method of bond formation that is different. The electrical charge of the cation, $NH_4^+$, is spread over the whole ion.

Dative covalent bonding can sometimes occur between molecules — for example, when ammonia combines with boron trifluoride, $BF_3$. The boron atom has six electrons in its valence shell and, therefore, requires two electrons to form an outer octet. These are provided by the lone pair of electrons on the nitrogen atom of the ammonia molecule:

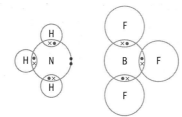

The lone pair of electrons donated by the nitrogen atom is shared with the boron atom. A dative covalent bond is formed and the nitrogen and boron atoms both satisfy the octet rule:

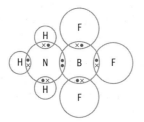

## Typical properties of covalent compounds

Many covalent compounds are liquids or gases with low melting points. If solid, they are not dense and tend to be soft and easily broken, unlike ionic crystals that are hard and have a specific shape. Candle wax, which is a mixture of solid covalent compounds, is an example.

Covalent compounds are poor conductors of electricity. These two properties can be explained by the nature of the covalent bond.

Simple covalent compounds have low melting points because they exist as discrete (separate) molecules and the attraction between neighbouring covalent molecules is weak. The attraction between molecules is due to **intermolecular forces**. Since the intermolecular forces are weak, only a small amount of energy is required to break them, so the melting points are low.

Covalent compounds do not conduct electricity because this requires mobile charge carriers (free electrons or free ions). Most covalent compounds do not have mobile charge carriers and hence are poor conductors.

# Shapes of simple molecules and ions

Simple covalent compounds exist as discrete molecules with atoms held together by shared pairs of electrons. Within these molecules, there may be two types of electron pair:

- A **bonded pair of electrons** is a pair of electrons shared between two adjacent atoms. The electrons move between the two atoms and are **delocalised** between them.

The expression 'delocalised' is used widely in chemistry. Electrons that are delocalised are not attached to one atom only; they can move to another atom. In a single bond, there is limited delocalisation because the electrons can move no further than that particular bond. Sometimes electrons have a greater degree of freedom — for example, in metallic bonding (pages 87–88).

- A **non-bonded pair of electrons** is a pair of electrons that is not involved in bonding. It is sometimes referred to as a lone pair. The electrons are not shared and are located on one atom only.

## Electron pair repulsion theory

It is the numbers of bonded and non-bonded electron pairs that determine the shape of a molecule. A simple model of molecular shape is provided by the electron pair repulsion theory. This is based on the fact that pairs of electrons (both bonded and non-bonded) repel each other to be as far apart in space as possible. This determines the basic shape of the molecule. However, the shape is often modified because non-bonded (lone) pairs repel more than bonded pairs.

Non-bonded electrons are slightly closer to the nucleus of an atom than the bonded pairs. This means that the non-bonded pairs of electrons are slightly nearer to neighbouring electrons. Therefore, they exert a stronger repulsive force, which modifies the shape of the molecule.

The key point is that electron pairs repel each other to be as far apart as possible. Lone pairs repel more than bonded pairs. From strongest to weakest, the order of strength of repulsion is:

lone pair–lone pair > lone pair–bonded pair > bonded pair–bonded pair

The overall shape of a covalent molecule (or ion) is determined by the number and type of electron pairs around the central atom.

The way to determine the shape is to draw a dot-and-cross diagram and use this to deduce the number (and types) of electron pairs around the central atom.

### Two bonded pairs and no lone pairs

Covalent molecules containing two single covalent bonds are not common. The simplest example is beryllium chloride, $BeCl_2$, which, unusually for a metallic compound, is covalent. The beryllium atom has two bonded pairs and no lone pairs. The bonded pairs repel each other until they are as far apart as possible:

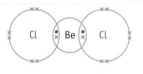

The resultant shape is linear; the bond angle is $180°$:

Linear shape is more common in molecules containing multiple — double or triple — bonds, such as carbon dioxide, $CO_2$. The central carbon atom and each oxygen atom in carbon dioxide are connected by two pairs of bonded electrons, each constituting a double covalent bond. The double-bonded pairs repel each other until they are as far apart as possible. The resultant shape is linear and the bond angle is $180°$:

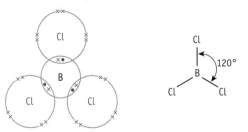

### Three bonded pairs and no lone pairs

Boron forms covalent molecules with the halogens. In boron trichloride, $BCl_3$, the central boron atom has three bonded pairs and no lone pairs. The bonded pairs repel each other so that they are as far apart as possible. The resultant shape is **trigonal planar**; the bond angle is $120°$:

## Four bonded pairs and no lone pairs

Carbon forms many compounds containing four covalent bonds. The simplest is methane. The central carbon atom in methane has four bonded pairs and no lone pairs. The bonded pairs repel each other so that they are as far apart as possible. The resultant shape is **tetrahedral**; the bond angle is 109.5°:

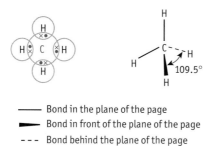

——— Bond in the plane of the page
⬤ Bond in front of the plane of the page
- - - Bond behind the plane of the page

## Five bonded pairs and no lone pairs

Elements in period 2 obey the octet rule. However, some elements in period 3 can form compounds in which there are more than eight electrons around the central atom. An example is phosphorus pentafluoride, $PF_5$. The phosphorus atom has five bonded pairs and no lone pairs. The bonded pairs repel each other so that they are as far apart as possible. The resultant shape is **trigonal bipyramidal** (base-to-base triangular pyramids; the bond angles are 120° and 90°:

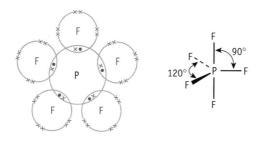

## Six bonded pairs and no lone pairs

Sulphur can also exceed the octet rule. Sulphur forms six covalent bonds in compounds such as sulphur hexafluoride. The sulphur atom has six bonded pairs and no lone pairs. The bonded pairs repel each other to the point of maximum separation to give an **octahedral** shape (base-to-base square pyramids); the bond angles are all 90°:

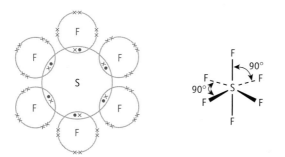

### Covalent molecules that contain lone pairs of electrons

Many covalent compounds contain lone pairs of electrons as well as bonded pairs. These lone pairs affect the shape of the molecule, because lone pairs repel more than bonded pairs.

#### Ammonia

A dot-and-cross diagram of ammonia is shown below:

The central nitrogen atom in ammonia has four pairs of electrons around it and, therefore, you might expect them to be repelled to form tetrahedral bond angles of 109.5°. However, the four pairs of electrons are not identical; there are three bonded pairs and one lone pair. The lone pair repels more than the bonded pairs, which results in the bonded pairs being forced closer together. The angle between the N–H bonds contracts from 109.5° to about 107°. The shape of the ammonia molecule is **pyramidal**:

#### Water

A dot-and-cross diagram for water is shown below:

The central oxygen atom in a water molecule has four pairs of electrons around it and, therefore, you might expect the molecule to be tetrahedral with a bond angle of 109.5°. However, the four pairs of electrons surrounding the oxygen atom consist of two bonded pairs and two lone pairs. The lone pairs repel more than the bonded pairs, forcing the bonded pairs closer together. The bond angle contracts from 109.5° to about 104.5°. The shape of the water molecule is **angular** or **bent**:

## Shapes of ions

Ions such as the ammonium ion, $NH_4^+$, and the amide ion, $NH_2^-$, contain covalent bonds and, therefore, have a fixed shape and a defined bond angle.

- When an ammonium ion is formed from ammonia, a proton is gained. The central nitrogen atom has four bonded pairs of electrons. The shape is tetrahedral and the bond angle is 109.5°.

■ When an amide ion is formed from ammonia, a proton is lost. The central nitrogen atom has two bonded pairs of electrons and two lone pairs. The shape is angular (bent) and the bond angle is about 104°.

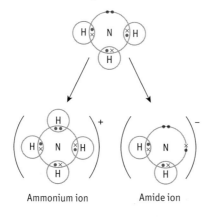

Ammonium ion      Amide ion

## Question

Determine the shape of each of the following molecules and ions. You may find it helpful to refer to the periodic table (page 264) to establish the number of outer-shell electrons.

**a** Hydrogen sulphide, $H_2S$
**b** Phosphine, $PH_3$
**c** Sulphur dichloride, $SCl_2$
**d** Dichloromethane, $CH_2Cl_2$
**e** $CH_3^+$
**f** $H_3O^+$

# Metallic bonding

The metallic 'bond' is the electrostatic attraction between the delocalised electrons and the positive ions held within the lattice.

Metal atoms pack together in a regular arrangement. The outer orbital electron clouds overlap and the electrons are free to move from one atom to another — they are **delocalised**. The electrons move through a giant lattice of positive ions. The delocalised electrons are sometimes called 'a sea of free electrons':

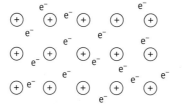

The '+' sign in this diagram indicates a positive metal ion, *not* a proton.

The delocalised electrons are free to move anywhere within the lattice, so the metallic bond is non-directional.

## Typical properties of metals

Metals (apart from mercury) are solids, most of which have high melting points. They are good conductors of electricity. These two properties can be explained by the nature of the metallic bond.

High temperatures are needed to melt metals because the metallic bond is strong and exists throughout the giant lattice. In order to melt a solid metal, all the bonds in the lattice have to be broken, which requires a large amount of energy.

Metals are good conductors of electricity because they contain mobile charge carriers that are free to move, even in the solid. The mobile charge carriers are the delocalised electrons.

Metals are ductile (can be stretched to form wire) and malleable (can be re-shaped). This is because the metallic bond is non-directional, is independent of shape and exists in all directions.

# Bond polarity and polarisation

It is rare for a compound to be 100% ionic. Similarly, other than simple diatomic molecules in which the covalent bond exists between two identical atoms, it is unusual to find a molecule that is 100% covalent. There is a range of intermediate types of bond between the extremes of ionic and covalent.

Many bonds are described as essentially:
- covalent with some ionic character
- ionic with some covalent character

## Bond polarity

In a covalent molecule, ionic character is introduced when the pair of electrons making up the covalent bond is not shared evenly between the component atoms. This is a result of the **electronegativity** of the atoms.

Electronegativity is unusual in that there is no consistent quantitative definition. It is sufficient to interpret it as the attraction of an atom, within a molecule, for the pair of electrons in a covalent bond. Table 6.2 is an attempt to quantify the effect by giving each atom a number to represent the strength of its attraction. However, such figures are not wholly reliable.

You should appreciate that:
- electronegativity increases across a period
- electronegativity decreases down a group

| Element | Electronegativity |
|---------|-------------------|
| H | 2.1 |
| Li | 1.0 |
| Na | 0.9 |
| K | 0.8 |
| Mg | 1.2 |
| Al | 1.5 |
| Si | 1.8 |
| P | 2.1 |
| S | 2.5 |
| Cl | 3.0 |
| Br | 2.8 |
| I | 2.5 |
| B | 2.0 |
| C | 2.5 |
| N | 3.0 |
| O | 3.5 |
| F | 4.0 |

*Table 6.2*
*Electronegativity*

In a covalent bond, the element with the greater electronegativity attracts electrons towards itself.

In a molecule of hydrogen, $H_2$, both hydrogen atoms have the same attraction for the two bonded electrons. Hence, they are shared equally. Molecules like hydrogen, $H_2$, are regarded as 100% covalent:

$$H \overline{\qquad} H$$
$$2.1 \qquad 2.1$$

In a molecule of hydrogen chloride, HCl, the hydrogen atom and the chlorine atom have a different attraction for the two bonded electrons. Hence, the electrons are not shared equally. The molecules are not 100% covalent because the electrons are more concentrated towards the chlorine atom:

$$H \longrightarrow Cl$$
$$2.1 \qquad 3.0$$

The chlorine atom becomes slightly negatively charged with respect to the hydrogen; the hydrogen is left with a slight positive charge. A small degree of charge is indicated on a molecule by using the symbol $\delta+$ or $\delta-$:

$$\delta+ \qquad \delta-$$
$$H \overline{\qquad} Cl$$

A covalent bond with some charge separation is said to have a **dipole**. A dipole indicates that the bond has a small amount of ionic charge; the bond is best described as essentially covalent, with some ionic character.

The effect that a difference in electronegativity has on a bond is shown in Table 6.3. The values given for the 'percentage ionic character' are approximations.

| Difference in electronegativity | Percentage ionic character |
|:---:|:---:|
| 0.3 | 2 |
| 0.4 | 4 |
| 0.5 | 6 |
| 0.6 | 9 |
| 0.7 | 12 |
| 0.8 | 15 |
| 0.9 | 19 |
| 1.0 | 22 |
| 1.1 | 26 |
| 1.2 | 30 |
| 1.3 | 34 |
| 1.4 | 39 |
| 1.5 | 43 |
| 1.6 | 47 |
| 1.7 | 51 |
| 1.8 | 55 |
| 1.9 | 59 |

*Table 6.3*
*Percentage ionic character*

The difference in electronegativity between hydrogen and chlorine is 0.9. This means that the bond between hydrogen and chlorine is essentially covalent (81%) with some ionic character (19%). This type of bond is described as a **polar bond**.

### Polar bonds and polar molecules

A polar bond is due to the difference in electronegativity of the atoms in a covalent bond. However, molecules with polar bonds are not necessarily polar molecules. If a molecule is symmetrical, the dipoles cancel out, even if the bonds are polar. For example, in a molecule of carbon dioxide, the electrons in the covalent bonds are pulled towards the oxygen atoms:

$$\overset{\delta-}{\underset{3.5}{O}} = \overset{\delta+}{\underset{2.5}{C}} = \overset{\delta-}{\underset{3.5}{O}}$$

However, because the molecule is symmetrical, the dipoles cancel each other out. The net result is that the $C=O$ bonds are polar, but the carbon dioxide molecule is not.

Symmetrical shapes include linear, trigonal planar, tetrahedral, trigonal bipyramidal and octahedral. If the atoms bonded to the central atom are the same, any dipoles in the bonds cancel each other out and the molecule is non-polar — for example, methane, $CH_4$.

Non-symmetrical shapes include pyramidal and angular (or bent). Molecules with these shapes are always polar, even when the central atom is attached to atoms that are the same. Ammonia molecules are pyramidal and polar; water molecules are angular and polar.

## Polarisation

So far, it has been sufficient to regard electrons as small, negatively charged particles. Now, we need to consider electrons in terms of orbitals. In an ionic compound, the positively charged ions (cations) attract the electrons of the negatively charged ions (anions). If this attraction is strong, it distorts the shape of the electron orbitals. The degree of distortion depends on the **charge density** of the cation:

$$\text{charge density} = \frac{\text{charge of the ion}}{\text{ionic radius}}$$

The distortion is greatest when the cation is small and highly charged and the anion is large and highly charged.

An example of a small, highly charged cation is $Al^{3+}$. It has a small radius because the loss of electrons results in a contraction of the aluminium atom. The radius of an aluminium atom is 0.125 nm, but the radius of the ion, $Al^{3+}$, is 0.045 nm. Therefore, $Al^{3+}$ has a high charge density.

An ion such as $O^{2-}$ is large because, although an oxygen atom has a small diameter, the addition of two electrons causes a noticeable increase in size. The radius of an oxygen atom is 0.066 nm, but the radius of the ion, $O^{2-}$, is 0.146 nm.

In aluminium oxide, $Al_2O_3$, the charge density of the $Al^{3+}$ is high and pulls electrons from the $O^{2-}$ ion to such an extent that the electrons are partly shared. This is known as **polarisation**. The electron orbitals in the $O^{2-}$ ion are distorted and pulled towards the $Al^{3+}$ ion:

The bonds in aluminium oxide are due to the electrostatic attraction between the $Al^{3+}$ and the $O^{2-}$ ions. The bonding is essentially ionic, but because of polarisation, there is some covalent character.

## Overview

A compound consisting of two or more different non-metals is essentially covalent with some ionic character. In general, the molecules of such compounds are polar. However, if the molecule is symmetrical this is not the case, because the dipoles cancel each other out.

A compound consisting of a metal and a non-metal is essentially ionic with some covalent character.

# Intermolecular forces

Ionic, covalent and metallic bonds are all strong bonds; to break them requires between $200\,kJ\,mol^{-1}$ and $600\,kJ\,mol^{-1}$. The types of interaction collectively called **intermolecular forces** are weak; to overcome them requires between $2\,kJ\,mol^{-1}$ and $40\,kJ\,mol^{-1}$.

## Van der Waals forces

The weakest of the intermolecular forces are **van der Waals forces**. These act between all molecules, polar or non-polar. To understand van der Waals forces you must appreciate that the electrons in a covalent bond are not static. Although it is helpful to think of a covalent bond in terms of a pair of electrons considered as particles, a bond is created by the linking of electron orbitals. Orbitals are not static. They can be thought of as pulsating volumes of electronic charge, the charge density of which is changing constantly. The result of these variations in electron density is that one part of the bond becomes more negative than another part. This creates an **oscillating** or **instantaneous dipole**. This, in turn, induces a dipole in neighbouring molecules. The attraction between these induced dipoles produces the weak intermolecular interactions called van der Waals forces or **induced dipole–dipole** interactions. The strength of the van der Waals forces depends on the number of electrons in the molecule. The greater the number of electrons in an atom or molecule, the greater are the van der Waals forces.

Although van der Waals forces are weak, their presence explains some physical properties. For example, the boiling point of butane ($C_4H_{10}$) is higher than that of propane ($C_3H_8$). This is because a molecule of butane has more bonds, and hence more electrons, than a molecule of propane. Therefore, the van der Waals forces are greater in butane, and it is these intermolecular forces that have to be overcome for boiling to occur.

It is van der Waals forces that are responsible for the increase in melting and boiling points as the number of carbon atoms increases within a homologous series of organic compounds (page 124).

## Permanent dipole–dipole interactions

A stronger force exists in **permanent dipole–dipole interactions** between polar molecules that are essentially covalent but have some ionic character. As a result of electronegativity differences, the dipole is always present.

This is illustrated using an organic molecule, chloromethane, $CH_3Cl$, as an example. The C–Cl bond has a dipole:

$$
\begin{array}{c}
\text{H} \\
| \\
\text{H} - \overset{\displaystyle}{\underset{\displaystyle |}{\text{C}}} \overset{\delta+}{\phantom{|}} - \overset{\delta-}{\text{Cl}} \qquad \text{or} \qquad \overset{\delta+}{H_3C} - \overset{\delta-}{Cl} \\
\text{H}
\end{array}
$$

Oppositely charged dipoles attract:

$$
\overset{\delta+}{H_3C} - \overset{\delta-}{Cl} \ -----\  \overset{\delta+}{H_3C} - \overset{\delta-}{Cl}
$$

Permanent dipole–dipole interactions are not strong, but they have more effect than the weak van der Waals forces that are also present.

## Hydrogen bonding

**Hydrogen bonding** is another type of intermolecular attraction. Hydrogen bonds exist between molecules containing hydrogen that is bonded to one of the electronegative elements, nitrogen, oxygen or fluorine. A hydrogen bond is a strong intermolecular attraction between molecules that involves a lone pair of electrons from the nitrogen, oxygen or fluorine.

Hydrogen bonds exist between molecules in ammonia, $NH_3$, water, $H_2O$, and hydrogen fluoride, HF. They are also present in compounds that contain the –OH group (e.g. alcohols and carboxylic acids) and compounds that contain the –$NH_2$ group (e.g. amines). Hydrogen bonding is important in amino acids, peptides and proteins.

Water can be used to illustrate some features of hydrogen bonding. The hydrogen bond is formed between a lone pair of electrons on the oxygen atom in one water molecule and a hydrogen atom in an adjacent water molecule:

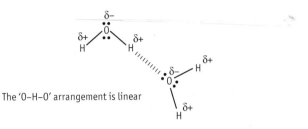

The 'O–H–O' arrangement is linear

The presence of a lone pair of electrons on the electronegative atom is essential for hydrogen bonding to take place.

Hydrogen bonding in water causes some anomalous properties. The most important are that:
- the solid (ice) is less dense than the liquid (water)
- water has unexpectedly high melting and boiling points

Ice floats on water because when water is frozen, the water molecules hydrogen-bond together to form interlocking hexagonal rings. This means that the molecules in any particular hexagon are held further apart, so that ice has a greater volume and thus lower density than water.

The structure of ice is complex, as can be seen in the patterns created by snowflakes.

SCOTT CAMAZINE/SPL

*Snowflakes have a characteristic six-fold symmetry and come in an infinite range of shapes*

Water can be regarded as a hydride of oxygen. If water followed the trend shown by the hydrides of the elements in same group of the periodic table (Figure 6.2), the melting point of water would be approximately $-100°C$ and the boiling point approximately $-80°C$.

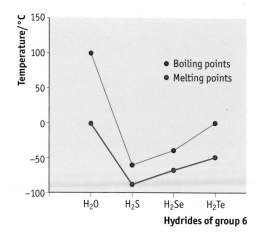

The unexpectedly high melting point of $0°C$ and boiling point of $100°C$ are due to the additional energy required to break the hydrogen bonds between adjacent water molecules. The intermolecular forces between the other group 6 hydrides are smaller and, therefore, less energy is required to break them.

# Structure and physical properties

There are many types of solid compound because of the different ways in which the atoms of elements or the molecules of compounds can combine. These differences depend on the sizes of atoms and the nature of the intermolecular forces. At a simple level, solids can be broadly classified as either simple molecular structures or giant lattice structures.

## Simple molecular lattices

Simple molecules do form larger structures. However, the forces between the molecules are weak and the solids melt easily. An example is solid iodine, $I_2$, which has a regular structure built up by repeating the section shown below:

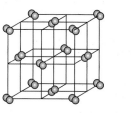

This basic unit is known as the **unit cell**. The crystal lattices is formed by repeating the unit cell.

# Giant structures

For details of giant ionic lattices, giant metallic lattices and their properties, see pages 77–78, 87 and 97.

### Giant covalent lattices

Giant covalent lattices are held together by strong covalent bonds between atoms. They have high melting points and boiling points. Apart from graphite, they have no mobile or free charged particles and do not conduct electricity. They are insoluble in polar and non-polar solvents because of the strong covalent bonds in the lattice structure.

Diamond and graphite both have giant covalent structures:
- In diamond, each carbon atom is bonded to four other carbon atoms.
- In graphite, each carbon atom is bonded to three other carbon atoms.

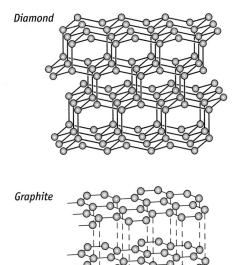

*Diamond*

*Graphite*

Diamond and graphite have different properties that are explained by their structures (Table 6.4).

|  | Diamond | Graphite |
|---|---|---|
| Structure | Tetrahedral, bond angle 109.5°; symmetrical structure held together by strong covalent bonds throughout lattice | Hexagonal layers, bond angle 120°; strong covalent bonds within layers. weak van der Waals forces between layers |
| Electrical conductivity | Poor conductor — no delocalised electrons; all outer-shell electrons used in covalent bonds | Good conductor — delocalised electrons move within the structure |
| Melting point | High melting point — strong covalent bonds throughout lattice | High melting point — strong covalent bonds within layers |
| Hardness | Hard — tetrahedral shape means external forces are spread evenly throughout the lattice | Soft — bonding within each layer is strong, but weak forces between the layers enable the layers to slide over each other |

*Table 6.4 Properties of diamond and graphite*

# Summary

You should now be able to describe:

- ionic bonding
- covalent bonding
- dative covalent (coordinate) bonding
- metallic bonding
- electronegativity and polar bonding
- intermolecular forces
- properties that relate to bonding and structure

# Additional reading

## 6

You may wonder how the crystal structures mentioned in this chapter are known. The overall shape of a large crystal can provide a clue — for example, a large sodium chloride crystal is cubic, which indicates that this is the way the ions are arranged. The discovery of X-rays, at the end of the nineteenth century, allowed a more reliable method of structure determination to be established. X-rays are part of the electromagnetic spectrum:

| Radiation | Wavelength/cm |
|-----------|---------------|
| Radio | $10^4$–$10^2$ |
| Microwave | 1 |
| Infrared | $10^{-2}$ |
| Visible | $10^{-5}$ |
| Ultraviolet | $10^{-6}$ |
| X-rays | $10^{-8}$ |
| Gamma rays | $10^{-10}$–$10^{-12}$ |

The wavelength of an X-ray is similar to the size of an atom and is much shorter than the wavelengths in visible light. X-rays have higher energy than visible light and are more penetrative. It is possible to use X-rays to investigate the structures of crystals. An intense beam of X-rays is focused onto the crystal and the X-rays are diffracted (dispersed). The diffraction pattern produced depends on the structure and orientation of the crystal. The detected pattern is then interpreted to provide information about the arrangement of particles within the crystal. The technique was developed at the Royal Institution in London by a father and son team, William and Lawrence Bragg. In 1915, they were awarded the Nobel prize in physics. Interpreting the data is not simple, but this is an important technique in the examination of materials.

Most metals have a simple repeating pattern that controls their overall geometry. This basic structural feature is called the unit cell and it defines how metal crystals are described. For example, one type of structure has a face-centred cube as the unit cell. A face-centred cube has atoms arranged at the corners of the cube with further atoms at the centre of each face. The faces are shared to generate the form of the metal crystal.

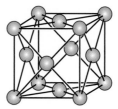

Aluminium, copper and gold have this structure, as does iodine. However, if you refer to the diagram on page 94, you will see that there are *molecules* of iodine, $I_2$, at each position of the face-centred cube.

The alkali metals are based on a unit cell that is a body-centred cube. The faces are shared with other unit cells to produce the overall structure. In a body-centred cubic structure, the atoms on the main diagonal are in contact with those at the corners. Metals are often described as having their atoms closely packed together. This is a fair description, but it is worth noting that a body-centred cubic structure is not as tightly packed as a face-centred cubic structure.

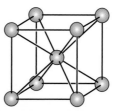

## Extension question: shapes of sulphur compounds

Electron pair repulsion theory can be used to predict the shapes of most molecules and ions. Sometimes, it requires careful consideration of how the electrons are distributed. An important point to note is that while atoms of the elements in the second period of the periodic table never have more than eight electrons in their outer orbits, those in the third and subsequent periods often do.

Sulphur forms two oxides, $SO_2$ and $SO_3$, and a range of anions, including sulphite, $SO_3^{2-}$, and sulphate, $SO_4^{2-}$. In addition, it forms thio compounds. In a molecule of a thio compound, an oxygen atom is replaced by a sulphur atom.

An example is thiosulphate, $S_2O_3^{2-}$.

a In the structures in parts (i) and (ii), the sulphur atom has ten electrons in its outer shell. What is the likely shape of:

(i) $SO_2$

(ii) $SO_3^{2-}$ (the two electrons forming the '2−' charge must be included in the structure; two of the oxygen atoms each have one extra electron added)

b In the sulphate ion, the sulphur atom accommodates twelve electrons in its outer orbit. What is the shape of:

(i) $SO_4^{2-}$

(ii) $S_2O_3^{2-}$

(In these two ions, two oxygen atoms each have one extra electron added.)

c Another feature of sulphur chemistry is that sulphur atoms can link together. Naturally occurring sulphur consists of $S_8$ molecules, in which the sulphur atoms are linked by single bonds to create a ring.

Draw an $S_8$ ring, giving some indication of its likely shape.

d There are various anions containing linked sulphur atoms. An example is the dithionite ion, $S_2O_4^{2-}$. This ion can be thought of as two $SO_2^-$ ions joined by two sulphur atoms to make the $^-O_2S-SO_2^-$.

Draw the electron structure of the $S_2O_4^{2-}$ ion and suggest its shape.

# Periodicity, group 2 and group 7

## Chapter 7

In this chapter some features of the periodic table are considered. Trends in the groups (vertical columns) and periods (horizontal rows) reflect the structures of the atoms of the elements within them, and these in turn affect the chemical properties of the elements. Similarities within groups are illustrated by considering the properties and chemical reactions of the elements in group 2 and group 7.

## The periodic table

During the first half of the nineteenth century, chemists built up knowledge of the behaviour of the elements and their compounds. This included their chemical properties, atomic masses and also the formulae of their compounds. It became clear that there were similarities between some elements and attempts were made to produce a simple classification to provide a guide to their properties. A breakthrough came when the Russian chemist Dmitri Mendeléev organised the elements according to their atomic masses and the formulae of their oxides and placed them in groups. This was the earliest version of the **periodic table**. Mendeléev knew nothing about electron structure, but what he did was to effectively arrange the elements in order of their increasing atomic number. This is the basis of the periodic table used today and it still useful as a means of classifying the elements. The key features of the periodic table are as follows:

- The vertical columns are called **groups** and the horizontal rows are called **periods**.
- Elements with the same outer-shell electron configuration are placed in the same group. For example, all elements with one electron (a single $s$-electron) in the outer shell are in group 1; all elements with four outer-shell electrons ($s^2 p^2$) are in group 4.
- The elements within a group have similar but steadily changing properties.
- Within a period, the elements go from metals on the left-hand side to non-metals on the right, with the noble gases on the extreme right.

In Table 7.1:
- the elements in red are metals
- the elements in blue are non-metals
- the elements in green show properties of both metals and non-metals and are called **metalloids**

Table 7.1
The periodic table

| | 1 | 2 | | | | | | | | | | | 3 | 4 | 5 | 6 | 7 | 0 |
|---|---|---|---|---|---|---|---|---|---|---|---|---|---|---|---|---|---|---|
| 1 | H | | | | | | | | | | | | | | | | | He |
| 2 | Li | Be | | | | | | | | | | | B | C | N | O | F | Ne |
| 3 | Na | Mg | | | | | | | | | | | Al | Si | P | S | Cl | Ar |
| 4 | K | Ca | Sc | Ti | V | Cr | Mn | Fe | Co | Ni | Cu | Zn | Ga | Ge | As | Se | Br | Kr |
| 5 | Rb | Sr | Y | Zr | Nb | Mo | Tc | Ru | Rh | Pd | Ag | Cd | In | Sn | Sb | Te | I | Xe |
| 6 | Cs | Ba | La* | Hf | Ta | W | Re | Os | Ir | Pt | Au | Hg | Tl | Pb | Bi | Po | At | Rn |
| 7 | Fr | Ra | Ac* | Rf | | | | | | | | | | | | | | |

Group 1 is a typical group. The elements in group 1 are known as the **alkali metals**. They have the following electronic structures:

- lithium, $_3$Li — $1s^2\ 2s^1$
- sodium, $_{11}$Na — $1s^2\ 2s^2\ 2p^6\ 3s^1$
- potassium, $_{19}$K — $1s^2\ 2s^2\ 2p^6\ 3s^2\ 3p^6\ 4s^1$
- rubidium, $_{37}$Rb — $1s^2\ 2s^2\ 2p^6\ 3s^2\ 3p^6\ 3d^{10}\ 4s^2\ 4p^6\ 5s^1$
- rubidium, $_{37}$Rb — $1s^2\ 2s^2\ 2p^6\ 3s^2\ 3p^6\ 3d^{10}\ 4s^2\ 4p^6\ 5s^1$
- caesium, $_{55}$Cs — $1s^2\ 2s^2\ 2p^6\ 3s^2\ 3p^6\ 3d^{10}\ 4s^2\ 4p^6\ 5s^2\ 4d^{10}\ 5p^6\ 6s^1$

Lithium is in the second period because the electron configuration of its outer shell is $2s^1$. The outer shell configuration of rubidium ($5s^1$) indicates that rubidium is the first element in the fifth period.

The way in which the elements are grouped means that there are trends in physical and chemical properties across periods and down groups.

## Trends across a period

Table 7.2 illustrates the trends across period 3.

*Table 7.2 Trends across period 3*

| | Sodium | Magnesium | Aluminium | Silicon | Phosphorus | Sulphur | Chlorine |
|---|---|---|---|---|---|---|---|
| Atomic radius/nm | 0.190 | 0.160 | 0.130 | 0.118 | 0.110 | 0.102 | 0.099 |
| Type of bonding | Metallic | Metallic | Metallic | Covalent | Covalent | Covalent | Covalent |
| Structure | Giant lattice | Giant lattice | Giant lattice | Giant lattice | Simple molecular | Simple molecular | Simple molecular |
| Electrical conductivity | Good | Good | Good | Poor | Poor | Poor | Poor |
| Melting point/°C | 98 | 649 | 660 | 1410 | 44* | 113 | −101 |
| Boiling point/°C | 883 | 1107 | 2467 | 2355 | 281 | 445 | −35 |
| First ionisation energy/kJ mol$^{-1}$ | 496 | 738 | 578 | 789 | 1012 | 1000 | 1251 |
| *White phosphorus | | | | | | | |

### Atomic radius

Atomic radius decreases across a period because the attraction between the nucleus and the outer electrons increases. Although the number of protons and the number of electrons both increase across a period, the protons are concentrated in a central position and exert a stronger pull on the more diffuse electrons, which results in the decrease in atomic radius.

### Bonding

Across a period, bonding changes from metallic to covalent. This means that the structure goes from giant lattice to simple molecular. Silicon has covalent bonding but also has a strong giant lattice similar to that of diamond (page 95).

### Electrical conductivity

Electrical conductivity is related to the type of bonding. Sodium, magnesium and aluminium have metallic bonding. Metallic elements are good conductors because they possess mobile, free outer-shell electrons. These electrons are delocalised, which allows metals to conduct heat and electricity, even in the solid state.

The remaining elements across period 3 are poor conductors because they do not contain any mobile free electrons.

Group 3 elements tend to be better conductors than elements in groups 1 and 2 because group 3 atoms have three electrons in their outer shells, compared with one and two outer-shell electrons for elements in groups 1 and 2 respectively.

### Melting point and boiling point

Melting points are related to the type of structure. Giant metallic and giant covalent lattices have strong bonds between the atoms throughout the lattice, all of which have to be broken. Boiling points then follow the same pattern at a higher temperature.

This requires a great deal of energy and hence the melting and boiling points are high. The converse is true for simple covalent molecules that have weak inter-molecular forces, such as van der Waals forces, since little energy is needed to overcome them. Therefore, the melting and boiling points are low.

### Ionisation energy

Ionisation energy shows a general increase across a period. This is discussed in detail on page 119.

## Trends down a group

The trends down group 1 and group 7 are shown in Tables 7.3 and 7.4 respectively.

| | Lithium | Sodium | Potassium | Rubidium | Caesium |
|---|---|---|---|---|---|
| Atomic radius/nm | 0.15 | 0.19 | 0.23 | 0.24 | 0.26 |
| Ionic radius/nm | 0.06 | 0.10 | 0.13 | 0.15 | 0.17 |
| Melting point/°C | 180 | 98 | 64 | 39 | 29 |
| Boiling point/°C | 1342 | 883 | 760 | 686 | 669 |
| First ionisation energy/kJ mol$^{-1}$ | 520 | 496 | 419 | 403 | 376 |

*Table 7.3 Trends down group 1*

**Table 7.4** *Trends down group 7*

|  | Fluorine | Chlorine | Bromine | Iodine |
|---|---|---|---|---|
| Atomic radius/nm | 0.07 | 0.10 | 0.11 | 0.13 |
| Ionic radius/nm | 0.14 | 0.18 | 0.20 | 0.22 |
| Melting point/°C | −220 | −101 | −7 | 114 |
| Boiling point/°C | −188 | −35 | 59 | 184 |
| First ionisation energy/kJ mol$^{-1}$ | 1681 | 1251 | 1140 | 1008 |

Atomic radius

Atomic radii increase down a group because the attraction between the nucleus and the outer electrons decreases. This is because the extra shells of electrons shield the outer electrons from the nucleus. Although the number of protons in the nucleus also increases, the extra shells and the extra shielding cause a decrease in effective nuclear charge. Therefore, the size of the atom increases down a group.

Melting point and boiling point

Melting and boiling points show a gradual change in both group 1 and group 7. In group 1, the melting and boiling points *decrease* down the group; in group 7 they *increase* down the group.

The trend is due to bonding and structure. In a metallic group, such as group 1, the melting (and boiling) points of the elements depend on the strength of the metallic bond. A metallic bond is defined as the attraction between the positive ions in the lattice and the free, delocalised electrons (page 87). The attraction depends on two factors:
- the greater the charge, the stronger the metallic bond
- the smaller the ionic radius, the stronger the metallic bond

All group 1 elements form ions with a 1+ charge. The ionic radius increases down the group (Table 7.3), so the strength of the metallic bond decreases down the group. Therefore, it follows that the melting and boiling points also decrease.

In a non-metallic group, such as group 7, the forces that have to be broken for both melting and boiling to occur are the intermolecular (van der Waals) forces (pages 91–92), which depend on the number of electrons. As group 7 is descended, the number of electrons increases, so the melting and boiling points also increase.

# Ionisation energy

The ionisation energy of an element was introduced in Chapter 1 as evidence for the existence of electron subshells. It is possible (although beyond the scope of this course) to measure the energy required to remove electrons from an atom in order to create positive ions. This can be done stepwise. If an atom has seven electrons, seven successive ionisation energies can be measured, as each electron is removed to create ions with a charge of up to 7+.

The first ionisation energy of an element is the energy required to remove one electron from the ground state of each atom in a mole of gaseous atoms of that element, to form a mole of gaseous ions of charge 1+.

This can be illustrated by an equation:

$$M(g) \rightarrow M^+(g) + e^-$$

The second ionisation energy is the energy required to remove one electron from each ion in a mole of gaseous ions of charge 1+, to form a mole of gaseous ions of charge 2+.

The third ionisation energy is the energy required to convert a mole of $M^{2+}(g)$ ions into $M^{3+}(g)$ ions, and so on. The $n^{th}$ ionisation energy refers to the process:

$$M^{(n-1)+}(g) \rightarrow M^{n+}(g) + e^-$$

As the ion produced becomes more positively charged, it gets more difficult to remove electrons; it resists the next electron being taken away.

### Successive ionisation energies of fluorine

The successive ionisation energies of fluorine (atomic number 9) are given in Table 7.5.

**Table 7.5** *Successive ionisation energies of fluorine/kJ mol$^{-1}$*

| 1st | 2nd | 3rd | 4th | 5th | 6th | 7th | 8th | 9th |
|---|---|---|---|---|---|---|---|---|
| 1680 | 3400 | 6000 | 8400 | 11000 | 15200 | 17900 | 92000 | 106000 |

The removal of the first electron from a mole of fluorine atoms requires an energy input of 1680 kJ. The $F^+(g)$ ion created contains nine protons and eight electrons. The protons in the ion hold the electrons more tightly than in the neutral atom. It is, therefore, not surprising that the second ionisation energy $(F^+(g) \rightarrow F^{2+}(g) + e^-)$ is greater, at 3400 kJ mol$^{-1}$.

It is tempting to think of this trend only in terms of the increasing positive charge. However, it is both an effect of the increasing nuclear charge and of the remaining electrons being pulled closer to the nucleus as a result of this increased charge. In other words, the radius of the positive ion becomes smaller as successive electrons are removed. Table 7.5 shows that the successive ionisation energies increase. The difference between the seventh and the eighth ionisation energies is far greater than expected from the previous ionisation energies. The reason is found by considering the electron structure of the fluorine atom: $1s^2\,2s^2\,2p_x^2\,2p_y^2\,2p_z^1$. The first seven electrons are removed from orbitals of quantum number 2 (the 2s- and the 2p-orbitals); the eighth electron is removed from the 1s-orbital.

To follow the sequence of events, consider the electron configurations at each stage:

- $F$ — $1s^2\,2s^2\,2p_x^2\,2p_y^2\,2p_z^1$
- $F^+$ — $1s^2\,2s^2\,2p_x^2\,2p_y^1\,2p_z^1$
- $F^{2+}$ — $1s^2\,2s^2\,2p_x^1\,2p_y^1\,2p_z^1$
- $F^{3+}$ — $1s^2\,2s^2\,2p_x^1\,2p_y^1$
- $F^{4+}$ — $1s^2\,2s^2\,2p_x^1$
- $F^{5+}$ — $1s^2\,2s^2$
- $F^{6+}$ — $1s^2\,2s^1$
- $F^{7+}$ — $1s^2$
- $F^{8+}$ — $1s^1$

**e** You must appreciate that ionisation energy always relates to the formation of a *positive* ion. Many elements normally form negative ions, but this is not what is being considered here.

The seventh ionisation energy is for the process:

$$F^{6+}(g) \rightarrow F^{7+}(g) + e^-$$

It requires the input of $17\,900\,kJ\,mol^{-1}$. At this stage, the second electron from the 2s-orbital is removed. The energy requirement, though substantial, is much less than the $92\,000\,kJ\,mol^{-1}$ required for the eighth ionisation. This removes the next electron, which is from the 1s-orbital:

$$F^{7+}(g) \rightarrow F^{8+}(g) + e^-$$

The reason is simple: the 1s-orbital has a considerably smaller radius, and is much closer to the nucleus, than the 2s- and 2p-orbitals. Since the attraction of the nucleus for the electrons depends on their distance from the nucleus, it follows that it is harder to remove an electron from the 1s-orbital than from a 2s- or 2p-orbital. This emphasises the general point that there is a considerable increase in the ionisation energy when an electron is removed from a shell closer to the nucleus.

Series of ionisation energies containing a large jump in the ionisation energy can be used to determine the probable outer electron configuration.

---

**Worked example 1**

The first eight ionisation energies ($kJ\,mol^{-1}$) of element X are:

    580, 1800, 2700, 11 600, 14 800, 18 400, 23 300 and 27 500

What is the likely arrangement of electrons in the outermost orbitals of an atom of element X?

**Answer**

There is a large jump ($2700\,kJ\,mol^{-1}$ to $11600\,kJ\,mol^{-1}$) between the third and fourth ionisation energies.

This suggests that in the fourth ionisation the electron is removed from a shell closer to the nucleus than in the third ionisation.

Therefore, it is likely that of an atom of X has three electrons in its outer orbitals, arranged $s^2\,p^1$.

---

The question in worked example 1 could have asked which group of the periodic table element X is found in. The answer is group 3, because all elements in this group have $s^2\,p^1$ as their outer electron configuration.

◀ The element is aluminium.

---

**Worked example 2**

The first ionisation energy of argon is $1520\,kJ\,mol^{-1}$; that of potassium is $420\,kJ\,mol^{-1}$. Give a reason for the difference in these energies.

**Answer**

In an atom of argon, the orbitals of quantum numbers 1, 2 and 3 are full. An atom of potassium has a further electron in the 4s-orbital. Since this orbital is further from the nucleus, it is easy to remove. This is confirmed by the lower first ionisation energy of potassium.

## Ionisation energies in group 1

The first ionisation energies (kJ mol$^{-1}$) of the elements in group 1 of the periodic table are shown in Table 7.6.

**Table 7.6** *First ionisation energies of the group 1 elements/kJ mol$^{-1}$*

| Lithium | Sodium | Potassium | Rubidium | Caesium |
|---------|--------|-----------|----------|---------|
| 520 | 500 | 420 | 400 | 380 |

As the group is descended, $s$-orbitals of increasing quantum number are occupied. For lithium, the outer orbital is $2s^1$, for sodium it is $3s^1$, for potassium it is $4s^1$ and so on. Although in each atom there are equal numbers of protons and electrons, the distance of the outermost electron from the nucleus increases down the group. The effect of this increase in orbital size is apparent in the decrease in the first ionisation energy.

Not only is the outermost electron further away from the nucleus in the larger atoms, there are also more electrons situated between this electron and the nucleus. A lithium atom has one pair of electrons in the $1s$-orbital between the nucleus and the outer $2s^1$ electron; a potassium atom has a complete set of $1s$-, $2s$-, $2p$-, $3s$- and $3p$-orbitals before the outermost $4s^1$-electron is reached. The presence of these electrons reduces the effectiveness of the nucleus in holding on to the outermost electron. This is called **electron shielding**. When an electron is shielded from the nucleus by inner electrons, it is easier to remove the most distant electron than if there were no shielding.

---

*Worked example*

Explain why the first ionisation energy of barium is lower than that of calcium.

**Answer**

Calcium and barium have the same outer electronic structure ($s^2 p^2$).

The barium atom has a larger radius than the calcium atom. Therefore, the outer electrons of barium are further away from the nucleus.

The outer electrons of barium are more shielded from the nucleus because there are more intermediate layers of electrons.

The force of attraction of the barium nucleus on the outermost electrons is lower than that in calcium.

Therefore, the first ionisation energy of barium is lower than that of calcium.

---

ⓔ Questions that ask for a comparison of ionisation energies are common in exams. Be careful that you consider all the relevant factors. Most questions of this type are worth 3 or 4 marks, so the examiner is expecting some detail.

## Questions

**1** For each of the following pairs, state which element has the higher first ionisation energy and explain your answer.
   **a** Mg and Al
   **b** Mg and Ca
   **c** Ne and Na

**2** Explain why the first ionisation energy of potassium is less than that of sodium.

**3** The following are the first seven successive ionisation energies of an element, M, measured in $kJ\,mol^{-1}$:

790, 1600, 3200, 4400, 16100, 19800, 23800

Suggest in which group of the periodic table you would expect to find element M. Explain your reasoning.

# Types of reaction

The way in which elements or compounds react can be categorised as one of a limited number of reaction types.

## Precipitation reactions

In a **precipitation reaction**, a cation and an anion in solution combine to form an insoluble substance that precipitates out as a solid.

Precipitation always occurs rapidly. For example, some carbonates are insoluble in water. If two solutions are mixed, one being an aqueous solution of a carbonate and the other containing a metal with an insoluble carbonate, the metal carbonate precipitates instantly. Insoluble copper(II) carbonate, for example, can be formed by mixing aqueous sodium carbonate with aqueous copper(II) sulphate:

$$Na_2CO_3(aq) + CuSO_4(aq) \rightarrow CuCO_3(s) + Na_2SO_4(aq)$$

The best way to summarise this is to use an ionic equation (Chapter 2, pages 30–34):

$$CO_3{}^{2-}(aq) + Cu^{2+}(aq) \rightarrow CuCO_3(s)$$

Precipitation reactions are used as a test for halide ions (pages 117–18).

## Redox reactions

In a **redox reaction**, electrons are transferred from one substance to another. Some everyday examples of redox reactions include combustion of fuels and the rusting of iron.

Oxidation was originally defined as the gaining of oxygen. For example, when zinc reacts with oxygen to form zinc oxide, the zinc is oxidised:

$$Zn(s) + O_2(g) \rightarrow ZnO(s)$$

Closer inspection shows that the zinc atom becomes a zinc ion and in doing so loses two electrons:

$$Zn \rightarrow Zn^{2+} + 2e^-$$

Therefore, a species is oxidised if it loses electrons; a species is reduced if it gains electrons (OILRIG):

**O**xidation
**I**s
**L**oss
**R**eduction
**I**s
**G**ain

## Oxidation number

Oxidation number is a convenient way of quickly identifying whether or not a substance has undergone oxidation or reduction. In order to work out the oxidation number, you must first learn a few simple rules (Table 7.7).

*Table 7.7 Rules for determining oxidation numbers*

| Rule | Example |
|------|---------|
| All elements in their natural state have the oxidation number zero | Hydrogen, $H_2$; oxidation number $= 0$ |
| Oxidation numbers of the atoms of any molecule add up to zero | Water, $H_2O$; sum of oxidation numbers $= 0$ |
| Oxidation numbers of the components of any ion add up to the charge on that ion | Sulphate, $SO_4^{2-}$; sum of oxidation numbers $= -2$ |

The oxidation numbers of some elements in a molecule or ion never change; other elements have more than one oxidation number.

When calculating the oxidation numbers of elements in either a molecule or an ion, you should apply the following order of priority:

1 The oxidation numbers of elements in groups 1, 2 and 3 are always +1, +2 and +3 respectively.
2 The oxidation number of fluorine is always −1.
3 The oxidation number of hydrogen is usually +1.
4 The oxidation number of oxygen is usually −2.
5 The oxidation number of chlorine is usually −1.

Using these rules, it is possible to deduce any oxidation number. However, it is important that they are applied *in sequence*, as hydrogen (−1 in metal hydrides) and oxygen (−1 in peroxides) can vary. The following worked examples show how the oxidation number of chlorine can also vary.

---

**Worked example 1**

Deduce the oxidation number of Cl in NaCl.

**Answer**

The sum of the oxidation numbers in NaCl must add up to zero.
Using the order of priority given above, the oxidation number of Na must be +1.
The oxidation number of Cl must balance this.
Therefore, the oxidation state of Cl in NaCl is −1.

---

**Worked example 2**

Deduce the oxidation number of Cl in NaClO.

**Answer**

The sum of the oxidation numbers in NaClO must add up to zero.
In order of priority, Na comes first and its oxidation number must be +1; O is second and its oxidation number is −2.
For the oxidation numbers to add up to zero, the oxidation number of Cl in NaClO must be +1.

---

## Question

Deduce the oxidation numbers of each element in the following:

**a** $H_2O$, $NaOH$, $KNO_3$, $NH_3$, $N_2O$

**b** $SO_4^{2-}$, $CO_3^{2-}$, $NH_4^+$, $MnO_4^-$, $Cr_2O_7^{2-}$

Consider the following reaction:

$$Mg(s) + H_2O(g) \rightarrow MgO(s) + H_2(g)$$

This reaction occurs when magnesium is heated with steam. It is easy to see that magnesium is oxidised (it has gained oxygen) and that water is reduced (it has lost oxygen). The oxidation numbers for this reaction are shown below:

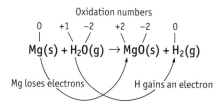

- An increase in oxidation number is due to oxidation.
- A decrease in oxidation number is due to reduction.

Electron transfer can be shown by half-equations. In this example, the half-equations are:

$$Mg \rightarrow Mg^{2+} + 2e^- \text{ (loss of electrons = oxidation)}$$
$$2H^+ + 2e^- \rightarrow H_2 \text{ (gain of electrons = reduction)}$$

- An **oxidising agent** encourages oxidation. It is a substance that receives electrons readily.
- A **reducing agent** encourages reduction. It is a substance that donates electrons readily.

Group 2 elements form 2+ ions readily, each atom releasing two electrons. Therefore, they are *reducing* agents.

Group 7 elements form 1− ions readily, each atom receiving one electron. Therefore, they are *oxidising* agents.

Electron loss occurs more readily on descending a group of the periodic table (page 101). It follows from this that, in group 2, barium is the best reducing agent. The opposite is true for the ability to gain an electron, so fluorine is the best oxidising agent.

The trends shown in Figure 7.1 occur elsewhere in the periodic table.

## Reactions of acids

Reactions of acids involve the transfer of a hydrogen ion from one substance (the acid) to another (the base). Since the hydrogen ion, $H^+$, is in fact a proton, acids are described as **proton donors**. These reactions are sometimes called **proton transfer** reactions.

Acids include:

- hydrochloric acid, $HCl$
- sulphuric acid, $H_2SO_4$
- nitric acid, $HNO_3$
- phosphoric acid, $H_3PO_4$
- ethanoic acid, $CH_3COOH$

| Group 2 | |
|---|---|
| Be | Reducing power increases *down* the group. Barium is the best reducing agent — a barium atom loses its outer electrons most easily. |
| Mg | |
| Ca | |
| Sr | |
| Ba | |

| Group 7 | |
|---|---|
| $F_2$ | Oxidising power increases *up* the group. Fluorine is the best oxidising agent — a fluorine atom gains an electron most readily. |
| $Cl_2$ | |
| $Br_2$ | |
| $I_2$ | |

*Figure 7.1 Reducing and oxidising power in groups 2 and 7*

In the absence of water, many of the substances called acids show none of the typical acid properties. It is unfortunate that it is usual to refer to solutions of acids without using the word 'solution'. We do, however, refer to dilute acids and this implies the presence of water. You should note that the state of the pure substances is sometimes unexpected. Hydrochloric acid, for example, is an aqueous solution of the covalent gas hydrogen chloride.

A **base** is a substance that can react with the hydrogen ion released by an acid. Bases can be regarded as **proton (hydrogen ion) acceptors**. Typical bases are oxides or hydroxides of metals.

If a base dissolves in water (most do not), it is referred to as an **alkali**. Examples include:

- sodium hydroxide, $NaOH$
- potassium hydroxide, $KOH$
- ammonia, $NH_3$

An alkali releases hydroxide ions ($OH^-$) in aqueous solution.

The inclusion of ammonia in the above list may seem odd because, as a gas, it does not contain hydroxide ions. However, when ammonia dissolves in water it reacts to produce the ionic compound ammonium hydroxide, $NH_4OH$. The reaction is an example of an **equilibrium reaction** since, once dissolved, there is a mixture of all four components, $NH_3$, $H_2O$, $NH_4^+$ and $OH^-$:

$$NH_3(g) + H_2O(l) \rightleftharpoons NH_4^+(aq) + OH^-(aq)$$

In this example it is easy to see why bases are regarded as proton acceptors. The ammonia molecule, $NH_3$, has received a hydrogen ion to become an ammonium ion, $NH_4^+$.

## Salts

Acids react with metals, carbonates, bases or alkalis to form **salts**.

> A salt is formed when an acid has one or more of its hydrogen ions replaced by either a metal ion or an ammonium ion.

The formation of salts is illustrated by the following typical reactions of acids.

### Reaction of an acid and a carbonate

Dilute hydrochloric acid reacts with sodium carbonate to produce sodium chloride, water and carbon dioxide, according to the equation:

$$2HCl(aq) + Na_2CO_3(aq) \rightarrow 2NaCl(aq) + H_2O(l) + CO_2(g)$$

The ionic equation is:

$$2H^+(aq) + CO_3^{2-}(aq) \rightarrow H_2O(l) + CO_2(g)$$

### Reaction of an acid and a base

Dilute hydrochloric acid reacts with magnesium oxide to produce magnesium chloride and water, according to the equation:

$$2HCl(aq) + MgO(s) \rightarrow MgCl_2(aq) + H_2O(l)$$

The ionic equation is:

$$2H^+(aq) + MgO(s) \rightarrow Mg^{2+}(aq) + H_2O(l)$$

### Reaction of an acid and an alkali

Dilute hydrochloric acid reacts with sodium hydroxide to produce sodium chloride and water, according to the equation:

$$HCl(aq) + NaOH(aq) \rightarrow NaCl(aq) + H_2O(l)$$

The ionic equation is:

$$H^+(aq) + OH^-(aq) \rightarrow H_2O(l)$$

Ammonia can react with an acid to produce a salt. Ammonium sulphate, $(NH_4)_2SO_4$, is used as a fertiliser. It is manufactured by reacting ammonia with sulphuric acid:

$$2NH_3(aq) + H_2SO_4(aq) \rightarrow (NH_4)_2SO_4(aq)$$

The ionic equation is:

$$2NH_3(aq) + 2H^+(aq) \rightarrow 2NH_4^+(aq)$$

### Reaction of an acid and a metal

Most metals react with an acid to produce a salt. Metals such as copper and silver react only with very concentrated acids. Some acids (e.g. nitric acid) produce complex products, but many acids (e.g. dilute hydrochloric and sulphuric acids) react with a metal to produce a salt and hydrogen.

Magnesium reacts with sulphuric acid to produce magnesium sulphate and hydrogen, according to the equation:

$$Mg(s) + H_2SO_4(aq) \rightarrow MgSO_4(aq) + H_2(g)$$

The ionic equation is:

$$Mg(s) + 2H^+(aq) \rightarrow Mg^{2+}(aq) + H_2(g)$$

It can be seen from the equations that this is a redox reaction. The magnesium atom loses two electrons to become an $Mg^{2+}$ ion; the two hydrogen ions from the acid each gain an electron to create a diatomic molecule of hydrogen gas. The magnesium is oxidised and the hydrogen ions are reduced.

In terms of oxidation number:
- the magnesium changes from oxidation number 0 to +2
- each hydrogen ion changes from oxidation number +1 to 0

All the reactions between acids and metals are redox reactions. A further illustration is the reaction of hydrochloric acid and zinc to produce zinc chloride and hydrogen:

$$Zn(s) + 2HCl(aq) \rightarrow ZnCl_2(aq) + H_2(g)$$

The ionic equation is:

$$Zn(s) + 2H^+(aq) \rightarrow Zn^{2+}(aq) + H_2(g)$$

The zinc is oxidised and the hydrogen is reduced.

### Anhydrous and hydrated salts

Most salts exist in the solid state either as pure substances or in the form of crystals that contain water molecules as part of their structure. The pure substance is **anhydrous**, which means that it contains no water. The crystals containing water are said to possess **water of crystallisation** and are, therefore, **hydrated**.

The water in hydrated salts is indicated by writing the formula of the substance followed by a full stop and the number of molecules of water. For example, iron(II) sulphate crystals have the formula $FeSO_4.7H_2O$.

Given the relationship between mass in grams and the amount of substance in moles (page 47), you should be able to use experimental data to determine the formula of a hydrated salt, as illustrated in the worked example below.

---

**Worked example 1**

A sample of copper sulphate crystals has a mass of 6.80 g. The sample is heated to drive off all the water of crystallisation. The mass is reduced to 4.35 g. Calculate the formula of the copper sulphate crystals.

**Answer**

mass of water in the crystals driven off by the heating = 6.80 − 4.35 = 2.45 g

amount (in moles) of water of crystallisation = $\dfrac{2.45}{18}$ = 0.136 mol

molar mass of anhydrous copper sulphate = 63.5 + 32.1 + (4 × 16.0) = 159.6 g

amount (in moles) of copper sulphate = $\dfrac{4.35}{159.6}$ = 0.02726 mol

ratio of moles of copper sulphate to moles of water = 0.02726:0.136 or 1:5

Therefore, the formula of hydrated copper sulphate is $CuSO_4.5H_2O$.

---

ⓔ Not all crystalline substances include water. Sodium chloride, for example, does not contain water of crystallisation.

> **Worked example 2**
>
> A sample of magnesium sulphate crystals contains 9.78% Mg, 38.69% $SO_4^{2-}$ and 51.53% $H_2O$ by mass.
>
> Determine the formula of the crystals.
>
> **Answer**
>
> amount (in moles) of magnesium $= \dfrac{9.78}{24.3} = 0.4025 \, mol$
>
> amount (in moles) of sulphate $= \dfrac{38.69}{96.1} = 0.4026 \, mol$
>
> amount (in moles) of water $= \dfrac{51.53}{18} = 2.863 \, mol$
>
> ratio $= 0.4025 : 0.4026 : 2.863$ or $1 : 1 : 7$
>
> Therefore, the formula of magnesium sulphate crystals is $MgSO_4.7H_2O$.

## Reactions that form complexes

Reactions that form complexes involve the attraction between a metal cation and an electronegative species with a spare pair of electrons. This type of reaction is not covered here, but you will meet such reactions if you study chemistry further.

# Group 2

## Electron configuration

Atoms of group 2 elements have two electrons in their outer shell and readily form 2+ ions (cations) that have the same electron configuration as a noble gas. The electron configurations of the elements are:

- beryllium, $_4Be$ — $1s^2 \, 2s^2$
- magnesium, $_{12}Mg$ — $1s^2 \, 2s^2 \, 2p^6 \, 3s^2$
- calcium, $_{20}Ca$ — $1s^2 \, 2s^2 \, 2p^6 \, 3s^2 \, 3p^6 \, 4s^2$
- strontium, $_{38}Sr$ — $1s^2 \, 2s^2 \, 2p^6 \, 3s^2 \, 3p^6 \, 3d^{10} \, 4s^2 \, 4p^6 \, 5s^2$
- barium, $_{56}Ba$ — $1s^2 \, 2s^2 \, 2p^6 \, 3s^2 \, 3p^6 \, 3d^{10} \, 4s^2 \, 4p^6 \, 4d^{10} \, 5s^2 \, 5p^6 \, 6s^2$

## Physical properties

The elements in group 2 are all metals. Therefore, they are good conductors of electricity and have high melting and boiling points. With the exception of beryllium, generally they form colourless ionic compounds that also have high melting and boiling points. Their compounds are good conductors of electricity when molten or in aqueous solution, but poor conductors when solid.

## Chemical properties

Atoms of group 2 elements react by losing two electrons:

$$M \rightarrow M^{2+} + 2e^-$$

**e** Questions are often set about conductivity. Remember that electricity is conducted in metals by free electrons; in ionic compounds that are molten or in aqueous solution, it is conducted by free ions.

Reactivity increases down the group. This is because the atomic radius increases down the group — hence the outer electrons are further from the nucleus and the intervening electron shells provide some shielding from the nucleus. These two factors outweigh the increased number of protons in the nucleus.

Reaction with oxygen

Group 2 elements undergo redox reactions with $O_2$:

- Magnesium burns, emitting a bright white light:

$$Mg(s) + \tfrac{1}{2}O_2(g) \rightarrow MgO(s)$$

- Calcium burns with a brick-red flame:

$$Ca(s) + \tfrac{1}{2}O_2(g) \rightarrow CaO(s)$$

- Strontium burns with a crimson flame:

$$Sr(s) + \tfrac{1}{2}O_2(g) \rightarrow SrO(s)$$

- Barium burns with a green flame:

$$Ba(s) + \tfrac{1}{2}O_2(g) \rightarrow BaO(s)$$

The oxidation number of the group 2 element increases from 0 to +2; the oxidation number of oxygen decreases from 0 to −2.

$$Mg(s) + \tfrac{1}{2}O_2(g) \rightarrow MgO(s)$$
$$0 \qquad\quad 0 \qquad\quad +2-2$$

*Group 2 compounds are important components of fireworks and help to produce a range of colours*

COREL

### Reaction with water

Group 2 elements undergo redox reactions with water to produce the hydroxide and hydrogen. Balanced equations for the reactions are:

$$Mg(s) + 2H_2O(l) \rightarrow Mg(OH)_2(s) + H_2(g)$$
$$Ca(s) + 2H_2O(l) \rightarrow Ca(OH)_2(aq) + H_2(g)$$
$$Sr(s) + 2H_2O(l) \rightarrow Sr(OH)_2(aq) + H_2(g)$$
$$Ba(s) + 2H_2O(l) \rightarrow Ba(OH)_2(aq) + H_2(g)$$

The oxidation number of the group 2 element increases from 0 to +2; the oxidation number of hydrogen changes from +1 to 0:

$$Mg(s) + 2H_2O(l) \rightarrow Mg(OH)_2(s) + H_2(g)$$
$$\phantom{Mg(s)}0 \quad\quad +1\,-2 \quad\quad +2\,-2\,+1 \quad\quad 0$$

The rate of reaction increases down the group, largely because of the ease of cation ($M^{2+}$) formation.

The reaction between magnesium and water is very slow. The resultant magnesium hydroxide, $Mg(OH)_2$, is barely soluble in water and forms a white suspension.

Magnesium reacts with steam to produce magnesium oxide and hydrogen:

$$Mg(s) + H_2O(g) \rightarrow MgO(s) + H_2(g)$$

### Reaction of group 2 oxides with water

All group 2 metal oxides react with water to form hydroxides:

$$MgO(s) + H_2O(l) \rightarrow Mg(OH)_2(s)$$
$$CaO(s) + H_2O(l) \rightarrow Ca(OH)_2(aq)$$
$$SrO(s) + H_2O(l) \rightarrow Sr(OH)_2(aq)$$
$$BaO(s) + H_2O(l) \rightarrow Ba(OH)_2(aq)$$

The reactions of group 2 oxides with water are *not* redox reactions. The oxidation numbers of all the elements remain the same, for example:

$$MgO(s) + H_2O(l) \rightarrow Mg(OH)_2(s)$$
$$+2\,-2 \quad\quad +1\,-2 \quad\quad +2\,-2\,+1$$

The resulting hydroxide solutions are alkaline and have pH values in the range 8–12. The pH depends on the concentration of the solution.

Calcium hydroxide is used in agriculture to neutralise acidic soils. Magnesium hydroxide is used as an antacid in some indigestion tablets.

### Thermal decomposition of group 2 carbonates

Group 2 carbonates decompose on heating to form oxides and carbon dioxide:

$$MgCO_3(s) \rightarrow MgO(s) + CO_2(g)$$
$$CaCO_3(s) \rightarrow CaO(s) + CO_2(g)$$
$$SrCO_3(s) \rightarrow SrO(s) + CO_2(g)$$
$$BaCO_3(s) \rightarrow BaO(s) + CO_2(g)$$

The ease with which the carbonates decompose decreases down the group. Beryllium carbonate, $BeCO_3$, is so unstable that it does not exist at room temperature. Barium carbonate, $BaCO_3$, requires strong heating to bring about the decomposition.

◀ Magnesium hydroxide is formed as a suspension.

◀ An aqueous solution of calcium hydroxide is known as **limewater**.

The thermal decomposition of group 2 carbonates is not a redox reaction. The oxidation number of all the elements remains the same, for example:

$$MgCO_3(s) \rightarrow MgO(s) + CO_2(g)$$

$\ \ \ $ +2 +4 −2 $\ \ \ \ \ $ +2 −2 $\ \ \ $ +4 −2

# Group 7

The elements in group 7 of the periodic table are called **halogens**.

## Electron configuration

Atoms of group 7 elements have seven electrons in their outer shell and readily form a 1− ion (anion) that has the same electron configuration as a noble gas. The electron configurations of the halogens are:

- fluorine, $_9F$ — $1s^2\ 2s^2\ 2p^5$
- chlorine, $_{17}Cl$ — $1s^2\ 2s^2\ 2p^6\ 3s^2\ 3p^5$
- bromine, $_{35}Br$ — $1s^2\ 2s^2\ 2p^6\ 3s^2\ 3p^6\ 3d^{10}\ 4s^2\ 4p^5$
- iodine, $_{53}I$ — $1s^2\ 2s^2\ 2p^6\ 3s^2\ 3p^6\ 3d^{10}\ 4s^2\ 4p^6\ 4d^{10}\ 5s^2\ 5p^5$

## Physical properties

- The elements in group 7 are non-metals and, hence, are poor conductors.
- At room temperature:
  - fluorine, $F_2$, is a yellow gas
  - chlorine, $Cl_2$, is a green gas
  - bromine, $Br_2$, is an orange–brown liquid
  - iodine, $I_2$, is a purple solid
- Descending the group from fluorine to iodine, there is an increase in van der Waals forces corresponding to the increased number of electrons in the molecules. This increase in van der Waals forces reduces volatility. Therefore, the melting and boiling points increase down the group.

## Chemical properties

The reactivity of the halogens decreases down the group. This is the opposite trend to the group 2 elements. Group 2 metals react by losing electrons and it becomes easier to lose them down the group (page 101). The halogens in group 7 react by *gaining* electrons to form halide anions. The ease with which the electron is gained by a halogen atom decreases down the group. This is because atomic radius and shielding increase down the group and this reduces the effective attraction of the nucleus for electrons.

Fluorine is the most reactive halogen. It readily gains electrons and is a powerful oxidising agent. Iodine is the least reactive halogen. In order of decreasing activity, halide ions are formed thus:

$$F_2 + 2e^- \rightarrow 2F^-$$
$$Cl_2 + 2e^- \rightarrow 2Cl^-$$
$$Br_2 + 2e^- \rightarrow 2Br^-$$
$$I_2 + 2e^- \rightarrow 2I^-$$

e Make sure that you know the difference between a halogen and a halide. Many students confuse chlorine with chloride.

## Displacement reactions

A halogen ($F_2$, $Cl_2$ or $Br_2$) displaces a heavier halide ($Cl^-$, $Br^-$ or $I^-$) from one of its salts (Table 7.8).

**Table 7.8**
*Displacement reactions*

| | Fluoride, $F^-$ | Chloride, $Cl^-$ | Bromide, $Br^-$ | Iodide, $I^-$ |
|---|---|---|---|---|
| Fluorine, $F_2$ | — | ✔ | ✔ | ✔ |
| Chlorine, $Cl_2$ | ✘ | — | ✔ | ✔ |
| Bromine, $Br_2$ | ✘ | ✘ | — | ✔ |
| Iodine, $I_2$ | ✘ | ✘ | ✘ | — |

- Fluorine displaces $Cl^-$, $Br^-$ and $I^-$ from solution.
- Chlorine displaces $Br^-$ and $I^-$.
- Bromine displaces $I^-$ only.
- Iodine cannot displace any of the above halides.

These displacement reactions illustrate the decrease in oxidising power down the group. Chlorine oxidises both bromide and iodide ions. When bromide ions are oxidised, orange–brown bromine is produced.

The oxidation reactions of chlorine are represented by the equations:

$$Cl_2 + 2Br^- \rightarrow 2Cl^- + Br_2$$
$$Cl_2 + 2I^- \rightarrow 2Cl^- + I_2$$

Bromine oxidises $I^-$ only. Iodine is formed, which in aqueous solution has a brown–black colour. Addition of an organic solvent produces a distinctive violet colour.

The oxidation of iodide ions by bromine is represented by the equation:

$$Br_2 + 2I^- \rightarrow 2Br^- + I_2$$

Iodine does not oxidise either chloride or bromide ions.

*Chlorine displaces bromide ions and bromine is produced*

CHARLES D. WINTERS/SPL

Each displacement reaction is a redox reaction. The halogen higher in the group gains electrons (is reduced) to form the corresponding halide:

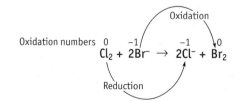

## Uses of chlorine

Chlorine is used in water treatment. It reacts reversibly with water and the resultant mixture kills bacteria.

$$Cl_2(aq) + H_2O(l) \rightleftharpoons HCl(aq) + HClO(aq)$$

This reaction is a redox reaction, but it is unusual in that chlorine undergoes both oxidation and reduction:

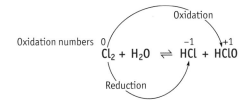

One chlorine atom in the $Cl_2$ molecule is oxidised. Its oxidation number changes from 0 to +1. The other chlorine atom is reduced. Its oxidation number changes from 0 to −1. This type of reaction is called **disproportionation.**

Chlorine reacts with sodium hydroxide to form bleach, which is a mixture of sodium chloride and sodium chlorate(I). This is also a disproportionation reaction of chlorine:

$$Cl_2(g) + NaOH(aq) \rightarrow NaCl(aq) + NaClO(aq)$$

## Testing for halide ions

Silver chloride, silver bromide and silver iodide are insoluble in water. Therefore the presence of chloride, bromide and iodide ions can be detected by the addition of an aqueous solution of silver nitrate, $AgNO_3(aq)$. The test solution is first acidified with dilute nitric acid and silver nitrate is then added. Each of the silver halides forms a different coloured precipitate. The identity of the precipitate is confirmed by its solubility in ammonia.

ⓔ The addition of dilute nitric acid prevents the formation of a precipitate of silver oxide, which would be misleading.

ANDREW LAMBERT PHOTOGRAPHY/SPL

*Precipitates of (a) silver chloride, (b) silver bromide and (c) silver iodide*

- Silver chloride is a white precipitate, soluble in dilute ammonia.

  $Ag^+(aq) + Cl^-(aq) \rightarrow AgCl(s)$
- Silver bromide is a cream precipitate, soluble in concentrated ammonia.

  $Ag^+(aq) + Br^-(aq) \rightarrow AgBr(s)$
- Silver iodide is a yellow precipitate, insoluble in concentrated ammonia.

  $Ag^+(aq) + I^-(aq) \rightarrow AgI(s)$

# Summary

You should now be able to:
- explain the structure of the periodic table in terms of periods and groups
- recognise where metals and non-metals are found in the periodic table
- state and explain the trends in atomic radius, structure and bonding, electrical conductivity, melting point and boiling point for the elements of period 3
- state and explain the trends in atomic radius, electrical conductivity, melting point and boiling point for the elements of groups 2 and 7
- understand what is meant by oxidation, reduction, oxidising agent and reducing agent
- calculate oxidation numbers for elements in compounds and ions
- describe the reactions of acids with carbonates, bases, alkalis and metals
- define ionisation energies
- understand the relationship between ionisation energy and the orbital structure of an element
- understand the trend in ionisation energies across a period of the periodic table
- explain the effect of atomic size on ionisation energy within a group of the periodic table
- describe the reactions of group 2 elements with oxygen and water and the trend in the ease of decomposition of their carbonates
- describe the elements of group 7 and their relative oxidising power with respect to their reactions with other halide ions
- describe some uses of chlorine
- describe the test for chloride, bromide and iodide ions

# Additional reading

## 7

The broad trends in ionisation energies have been discussed in this chapter. However, there are some subtle effects within these trends. Consider the successive ionisation energies (in kJ mol$^{-1}$) of nitrogen.

| 1st | 2nd | 3rd | 4th | 5th | 6th | 7th |
|------|------|------|------|------|------|------|
| 1400 | 2900 | 4600 | 7500 | 9400 | 53 300 | 64 300 |

The big increase in energy for the sixth ionisation (53 300 kJ mol$^{-1}$) can be readily understood because at this stage all electrons have been removed from the 2s- and 2p-orbitals and it is now necessary to remove an electron from the 1s-orbital.

However, there is a less obvious variation in the ionisation energies that might be overlooked. Examine the differences between the first five ionisation energies:

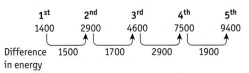

Between the third and fourth ionisation there is a jump in energy of 2900 kJ mol$^{-1}$, which is more than can be explained by the steady increase in the other values. 4600 kJ mol$^{-1}$ is required to remove a mole of electrons from the orbital structure $1s^2\,2s^2\,2p_x^{\,1}$ to give $1s^2\,2s^2$. The next ionisation involves the removal of one of the electrons from the 2s-orbital. This process requires more energy than expected because of the different shapes of the orbitals.

The 'dumb-bell' shape of the 2p-electron distribution protrudes beyond the more compact shape of the s-orbitals and this causes the 2p-electrons to be more shielded from the nucleus by the intervening 1s- and 2s-orbitals. This makes it easier to ionise an element if removing an electron from a p-orbital. However, if the electron has to be removed from the 2s-orbital, there is a jump in the energy required.

These factors can be used to explain the detailed shape of a graph of first ionisation energy of the elements against atomic number (Figure 7.2).

■ The big drop in the value between helium and lithium is because the electron configuration of helium is $1s^2$ and that of lithium is $1s^2\,2s^1$, so there is a noticeable expansion of size (as well as some degree of electron shielding).

■ It is easier to form an ion from sulphur than from oxygen because of a combination of the increase in size and the effect of electron shielding.

■ It is more difficult to form an ion from magnesium than it is from aluminium because magnesium has the outer electron configuration $3s^2$, whereas aluminium, with the structure $3s^2\,3p^1$, has a p-electron shielded by the 3s-electrons.

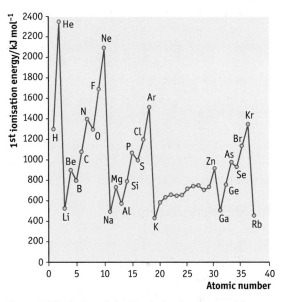

**Figure 7.2** *Variation of the first ionisation energy with atomic number*

The dip between nitrogen and oxygen is also explained by orbital configurations:

■ The ground state of oxygen is $1s^2\,2s^2\,2p_x^{\,2}\,2p_y^{\,1}\,2p_z^{\,1}$. On forming an ion it loses one of the pair of electrons in the $2p_x$-orbital.

■ The ground state of nitrogen is $1s^2\,2s^2\,2p_x^{\,1}\,2p_y^{\,1}\,2p_z^{\,1}$. On forming an ion it loses one of the unpaired 2p-electrons.

It is easier to cause two electrons to 'unpair' than it is to remove an unpaired electron from a given orbital because the paired electrons repel each other to some extent. Therefore, an ion is formed more easily from oxygen than from nitrogen. The same explanation accounts for the first ionisation values of:

■ phosphorus and sulphur
■ arsenic and selenium

Two elements are adjacent to each other in a row of the periodic table. Their successive ionisation energies are shown in Figure 7.3.

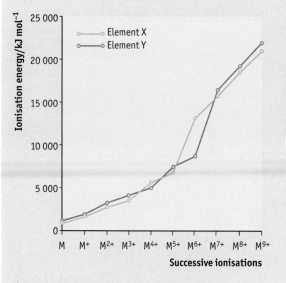

*Figure 7.3 Successive ionisation energies of two adjacent elements*

On the *x*-axis, each point represents the ionisation energy for that ion. So, for example, the point at $M^{4+}$ gives the value of the ionisation energy for the ionisation:

$$M^{4+}(g) \rightarrow M^{5+}(g) + e^-$$

a Decide in which groups of the periodic table elements X and Y would be found.

Element Y has an atomic mass of 127. Element X has several isotopes, as shown in the following table:

| Mass number | 120 | 122 | 123 | 124 | 125 | 126 | 128 | 130 |
|---|---|---|---|---|---|---|---|---|
| % abundance | 0.1 | 2.5 | 0.9 | 4.6 | 7.0 | 18.7 | 31.7 | 34.5 |

b Use this information to calculate the weighted mean atomic mass of element X.

c Compare your answers to a and b and suggest what is unusual about elements X and Y.

d Use the periodic table on page 264 to identify elements X and Y.

e Find one other place in the periodic table where the same pattern for two adjacent elements is observed.

# Unit 2
# Chains, energy and resources

# Basic concepts in organic chemistry

## Chapter 8

In this chapter you will learn about the variety of organic compounds, how they are bonded together and how they are named. Topics covered include:

- **isomerism** — some compounds have the same formula but different structures.
- **specialised vocabulary**
- **empirical, molecular, structural, displayed and skeletal formulae** — the distinction between the different types of formulae is explained and the importance of empirical formulae is discussed.
- **percentage yield** — a feature of organic reactions is that the actual yield is often less than the theoretical yield. Examples of calculations of percentage yield are given.
- **atom economy** — a recent consideration in a move towards 'greener' chemistry is the atom economy of reactions. The aim is to judge the efficiency of a chemical reaction by how much wastage there is in making the product. It is sensible to limit unnecessary side reactions where possible. The estimation of atom economy is illustrated with some simple calculations.

## Functional groups

Organic chemistry is the chemistry of carbon compounds. Carbon is unique in that an atom of carbon can form four strong covalent bonds and can bond with other carbon atoms to form chains and rings. The chains vary in length from two up to thousands of carbon atoms. This gives rise to millions of organic compounds. Carbon compounds range from the simple, for example methane, $CH_4$, to the complex, such as DNA. There is a huge range in between.

*Variety in organic molecules: (left) methane and (right) DNA*

RICHARDSON/CUSTOM MEDICAL STOCK PHOTO/SPL

Organic compounds always contain carbon and usually hydrogen. Some organic compounds also contain oxygen, nitrogen or a halogen, and a few contain sulphur.

As there are millions of organic compounds, the only way to study them is to group them together in families called **homologous series**. A homologous series is a group of compounds:

- that have the same general formula
- that contain the same functional group
- in which each member of the series differs from the next by $CH_2$

A **functional group** is either a structural feature (e.g. a carbon-to-carbon double bond, C=C), a group of atoms (e.g. a hydroxyl group, O–H) or a single atom (e.g. Cl). It is the functional group that determines much of the chemistry of a compound.

The AS course covers the chemistry of four homologous series — alkanes, alkenes, alcohols and halogenoalkanes. As you study the reactions of these compounds, you will also have to recognise other functional groups, such as those in aldehydes, ketones, carboxylic acids, esters and amines (Table 8.1).

ℯ In an exam you may be asked to define a homologous series or explain what is meant by a functional group.

*Table 8.1*
*Functional groups*

| Name | Functional group | General formula* | Prefix or suffix | First member | Second member† |
|---|---|---|---|---|---|
| Alkane | None — contains C–C and C–H single bonds | $C_nH_{2n+2}$ | -ane | $CH_4$ Methane | $C_2H_6$ Ethane |
| Alkene | C=C | $C_nH_{2n}$ $n \neq 1$ | -ene | $C_2H_4$ Ethene | $C_3H_6$ Propene |
| Alcohol | —C—OH | R–OH | -ol | $CH_3OH$ Methanol | $C_2H_5OH$ Ethanol |
| Halogenoalkane | —C—Cl | R–X | Chloro- Bromo- Iodo- | $CH_3Cl$ Chloromethane | $C_2H_5Cl$ Chloroethane |
| Aldehyde | —C(=O)H / R—C(=O)H | | -al | HCHO Methanal | $CH_3CHO$ Ethanal |
| Ketone | —C(=O)— / R—C(=O)R | | -one | $CH_3COCH_3$ Propanone | $C_2H_5COCH_3$ Butanone |
| Carboxylic acid | —C(=O)OH | R–COOH | -oic acid | HCOOH Methanoic acid | $CH_3COOH$ Ethanoic acid |

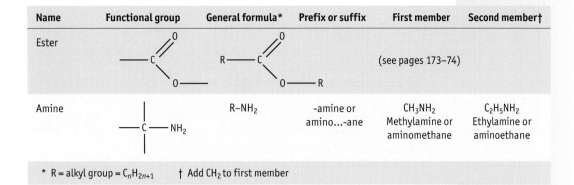

| Name | Functional group | General formula* | Prefix or suffix | First member | Second member† |
|---|---|---|---|---|---|
| Ester | | | (see pages 173–74) | | |
| Amine | | $R–NH_2$ | -amine or amino...-ane | $CH_3NH_2$ Methylamine or aminomethane | $C_2H_5NH_2$ Ethylamine or aminoethane |

\* R = alkyl group = $C_nH_{2n+1}$    † Add $CH_2$ to first member

# Naming organic compounds

With such a wide variety of compounds it is necessary to have an organised way of naming them. Nomenclature should follow the International Union of Pure and Applied Chemistry (IUPAC) rules. The IUPAC naming rules are based around the systematic names given to the alkanes and the prefix or suffix given to each functional group. The prefixes and suffixes for the functional groups are given in Table 8.1.

First, you need to learn the names of the alkanes (Table 8.2).

The name of the alkane is derived from the name of the longest continuous carbon chain. The alkyl, R, groups derived from the name of the alkanes are then used to name any side chains or branches (Table 8.3).

| Formula | Name |
|---|---|
| $CH_4$ | Methane |
| $C_2H_6$ | Ethane |
| $C_3H_8$ | Propane |
| $C_4H_{10}$ | Butane |
| $C_5H_{12}$ | Pentane |
| $C_6H_{14}$ | Hexane |
| $C_7H_{16}$ | Heptane |
| $C_8H_{18}$ | Octane |
| $C_9H_{20}$ | Nonane |
| $C_{10}H_{22}$ | Decane |

***Table 8.2** Formulae and names of the first ten alkanes*

| Number of carbon atoms in the longest straight chain | Stem name of longest straight chain | Number of carbon atoms in a side chain or branch | Name of side chain or branch |
|---|---|---|---|
| 1 | Meth- | 1 | Methyl |
| 2 | Eth- | 2 | Ethyl |
| 3 | Prop- | 3 | Propyl |
| 4 | But- | 4 | Butyl |
| 5 | Pent- | 5 | Pentyl |
| 6 | Hex- | 6 | Hexyl |
| 7 | Hept- | 7 | Heptyl |
| 8 | Oct- | 8 | Octyl |
| 9 | Non- | 9 | Nonyl |
| 10 | Dec- | 10 | Decyl |

***Table 8.3** Naming organic compounds*

### Worked example 1

Give the full name of the compound:

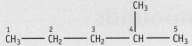

**Answer**

The longest carbon chain contains five atoms, so the name includes **pent-**.

This is an alkane, so it ends with **-ane**.

There is a branch containing one carbon atom, hence the name contains **methyl**.

It is possible to number the carbon chain from the left or the right.

Numbering from the left, the methyl group is on the fourth carbon atom – 4-methyl:

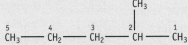

Numbering from the right, the methyl group is on the second carbon atom – 2-methyl.

The convention is to number from the chain end that results in the lowest number. In this case, numbering is from the right.

The full name of the compound is **2-methylpentane**.

### Worked example 2

Give the full name of the compound:

**Answer**

The longest carbon chain contains four atoms, so the name includes **but-**.

This is an alkene, so it ends with **-ene**.

The double bond starts on carbon atom 1, hence **-1-ene**.

The branch on the second carbon of the longest chain contains one carbon atom, hence **2-methyl**.

The full name is **2-methylbut-1-ene**.

### Worked example 3

Give the full name of the compound:

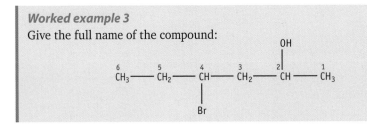

# Types of formula

Different types of formula are used in organic chemistry. You must be familiar with empirical and molecular formulae and be able to draw structural formulae, displayed formulae and skeletal formulae.

An **empirical formula** gives the simplest ratio of the elements in a compound. These can be calculated from the amounts of each element (see page 133).

A **molecular formula** represents the actual number of atoms of each element in the molecule. It does not provide any detail of the arrangement of the atoms. For example, the molecular formula of ethanol is $C_2H_6O$.

A **structural formula** uses the least amount of detail needed, using conventional groups, for an unambiguous structure. There are two possibilities for the structure of a compound of molecular formula $C_4H_{10}$: butane and methylpropane. The formula $C_4H_{10}$ is, therefore, ambiguous.
- The structural formula of butane is $CH_3CH_2CH_2CH_3$.
- The structural formula of methylpropane is $(CH_3)_3CH$.

The structural formula, therefore, makes it clearer how the atoms are bonded.

A **displayed formula** gives yet more information. It shows the relative positions of atoms and the number of bonds between them. Displayed formulae sometimes indicate the bond angles around each carbon atom. For butane and methylpropane the full displayed formulae are:

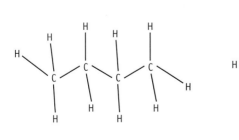

Butane

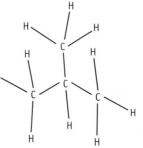

Methylpropane

Displayed formulae are usually simplified:

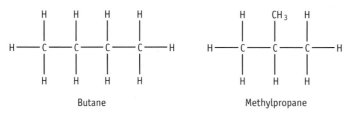

Butane                                    Methylpropane

The molecular formula of ethanol is $C_2H_6O$ and the structural formula is $CH_3CH_2OH$. The displayed formula is:

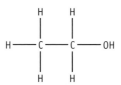

A **skeletal formula** is used to show a simplified organic structure by removing hydrogen atoms from alkyl chains, leaving just a carbon skeleton and associated functional groups. The skeletal formulae for butane and 2-methylpropane are shown below:

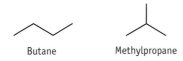

Butane                    Methylpropane

Some cyclic alkanes and benzene are represented as shown below:

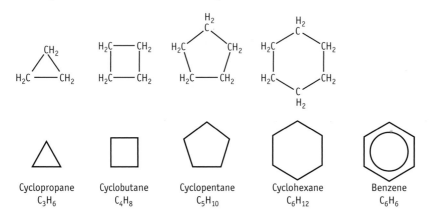

| Cyclopropane | Cyclobutane | Cyclopentane | Cyclohexane | Benzene |
| $C_3H_6$ | $C_4H_8$ | $C_5H_{10}$ | $C_6H_{12}$ | $C_6H_6$ |

# Isomerism

## Structural isomerism

**Structural isomers** are compounds that have the same molecular formula but different structural formulae. They can occur with all functional groups.

There are three types of structural isomer: chain, positional and functional group.

**e** You must be able to draw and name structural isomers.

## Chain isomerism

Chain isomers have differences in chain length. Examples include:

- butane and methylpropane

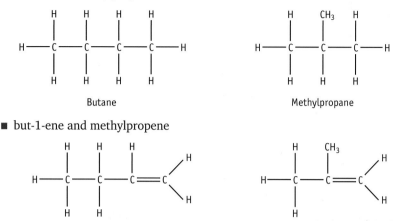

Butane                    Methylpropane

- but-1-ene and methylpropene

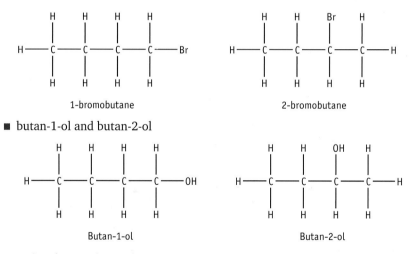

But-1-ene                    Methylpropene

## Positional isomerism

Positional isomers differ by the position of their functional group or groups. Examples include:

- 1-bromobutane and 2-bromobutane

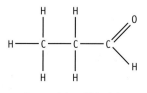

1-bromobutane                    2-bromobutane

- butan-1-ol and butan-2-ol

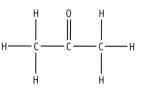

Butan-1-ol                    Butan-2-ol

## Functional-group isomerism

Functional-group isomers have different functional groups. An example is propanal (an aldehyde) and propanone (a ketone):

Propanal (an aldehyde)                    Propanone (a ketone)

# Stereoisomerism

Stereoisomers have the same molecular and structural formulae but a different three-dimensional spatial arrangement. An example of this is $E/Z$ isomerism.

## E/Z isomerism

$E/Z$ isomerism is found in some alkenes. The key features of $E/Z$ isomerism are:
- the presence of a C=C double bond
- each carbon atom in the C=C double bond is bonded to two different atoms or groups

As explained on page 159, the C=C double bond is rigid. Substituents on the molecule cannot twist around the double bond and move from one side to the other. This means that the following molecules are, in fact, different:

If the bond between the carbon atoms were single, the molecules would be the same, because a single bond allows the substituents on the carbon atoms to rotate freely. This is not the case with a double bond.

But-1-ene and but-2-ene both have a C=C double bond that prevents free rotation. However, one carbon atom in the C=C double bond in but-1-ene is bonded to two hydrogen atoms, so but-1-ene does not exhibit $E/Z$ isomerism. In but-2-ene, both carbon atoms in the double bond are bonded to different atoms/groups:

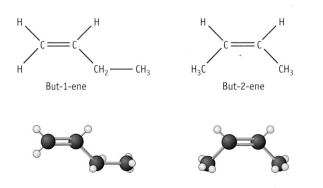

But-1-ene

But-2-ene

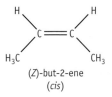

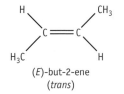

But-2-ene possesses both essential key features, and hence has a (Z)- and an (E)- isomer:

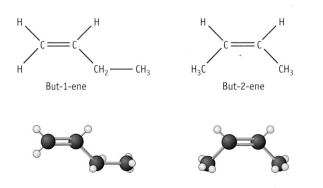

(Z)-but-2-ene
(cis)

(E)-but-2-ene
(trans)

**e** *E/Z* isomers are perhaps better known as *cis/trans* isomers, where *E* = *trans* and *Z* = *cis*. Both notations are used in this book.

**e** There is another type of stereoisomerism called optical isomerism. This is covered at A2.

◀ The prefixes '*Z*' and '*E*' come from the German words '*zusammen*', meaning together (i.e. on the same side of the molecule), and '*entgegen*', meaning opposite.

**e** In the exam, the (*Z*)-isomer will always be a *cis*-isomer and the (*E*)-isomer will always be the *trans*-isomer

## Summary of isomerism

- Structural isomers have the same molecular formula but different structures, for example:

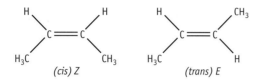

- *E/Z* isomers require a C=C double bond. The C=C prevents freedom of rotation. Each carbon atom in the double bond must be joined to two different atoms or groups, for example:

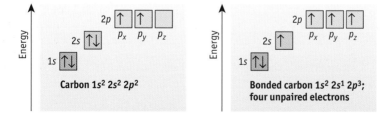

# Bonding in organic compounds

Organic compounds are essentially covalent. A covalent bond is a shared pair of electrons between two atoms. Each atom provides one electron of the shared pair.

All organic compounds contain carbon. The electron configuration of carbon, $_6$C, is $1s^2\, 2s^2\, 2p^2$. At first sight, it appears that only the two $2p$-electrons are available to form covalent bonds. However, when carbon forms bonds with other atoms, one of the $2s$-electrons is promoted to the third $2p$-orbital:

Carbon now has four unpaired electrons available for sharing and, therefore, can form four covalent bonds. The energy released by forming these bonds more than compensates for the small amount of energy required to promote one of the $2s$-electrons to the empty $2p$-orbital.

When a carbon atom bonds with another atom, either the $2s$- or one of the $2p$-orbitals is used. If the $s$- or $p$-orbital in carbon overlaps 'head-on' with an $s$- or $p$-orbital from another atom, in such a way that the overlap lies between the centres of the two atoms, a σ-**bond** (sigma-bond) is formed. If the $p$-orbital in carbon overlaps sideways with a $p$-orbital from another atom, a π-**bond** (pi-bond) is formed. The π-bond is discussed in more detail in the section on alkenes (page 158).

Note that the covalent bond in hydrogen is also an example of a σ-bond:

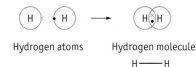

Hydrogen atoms     Hydrogen molecule

H———H

Alkanes consist of carbon and hydrogen only (e.g. methane, $CH_4$). The bonds are all σ-bonds. The four electron pairs involved mean that the bonds around the carbon atom point to the corners of a regular tetrahedron (page 85). The bond angle is 109.5°. This seems surprising when you consider that three of the orbitals used are *p*-orbitals that were originally at right angles to each other. However, when the electron is promoted from the 2*s*-orbital, some intermixing of the orbitals takes place. This is called **hybridisation**, and the shapes of the orbitals and the angles between them become symmetrical.

Carbon has an electronegativity of 2.5; hydrogen has an electronegativity of 2.1. Therefore, the C–H bond is slightly polar, with the electrons drawn towards the carbon atom:

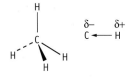

This applies to all four bonds. Therefore, in a molecule of methane the electrons in each C–H bond are drawn towards the central carbon atom:

Since the $CH_4$ molecule is symmetrical, the dipoles cancel each other out. Methane is non-polar, as are all alkanes.

Functional groups such as those in halogenoalkanes and alcohols disturb the alkane symmetry. This, together with the electronegativities of the halogen and oxygen, makes the molecules polar — for example, chloromethane and methanol:

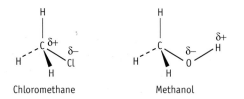

Chloromethane             Methanol

Non-polar molecules such as alkanes are unreactive compared with polar molecules such as chloromethane and methanol. The strength and the polarity of the bonds affect the chemical properties of a molecule.

◀ The hybridisation of orbitals is a common feature of bonding, but it is not considered further in this course and will not be examined.

# Important terms

In order to understand the chemistry of compounds containing functional groups, you need to know the following definitions:

- **Reagents**   These are the chemicals involved in a reaction.
- **Conditions**   These normally describe the temperature, pressure, solvent and the use of a catalyst.
- **Radical**   This is a reactive particle with an unpaired electron. The symbol for a radical shows the unpaired electron as a dot, e.g. $Cl\bullet$, $\bullet CH_3$.
- **Electrophile**   A reactive ion or molecule that attacks an electron-rich part of a molecule to form a new covalent bond. An electrophile is an **electron-pair acceptor**. Examples include $H^+$, $NO_2^+$ and $Br_2$.
- **Nucleophile**   An ion or molecule that attacks an electron-deficient part of a molecule. A nucleophile has a lone pair of electrons that can form a new covalent bond. A nucleophile is an **electron-pair donor**. Examples include $OH^-$ and $NH_3$.
- **Mechanism**   The overall reaction is broken down into separate steps. It is usual to identify the attacking species. This can be a radical, an electrophile or a nucleophile. The mechanism of a reaction usually involves the movement of electrons from one reagent to another. Curly arrows are used to represent the moving electrons:

> The curly arrow is used to show the movement of a pair of electrons

> The half-headed curly arrow is used to show the movement of a single unpaired electron

*e* For a reaction of any functional group you are expected to know the reagents and conditions and to be able to write a balanced equation and, often, the mechanism.

# Calculations in organic chemistry

Most chemistry exams contain at least one question that requires a calculation. Organic chemistry calculations can include the determination of:

- empirical and molecular formulae
- percentage yield
- atom economy

## Empirical and molecular formulae

The empirical formula is the simplest, whole-number ratio of atoms of each element in a compound. The molecular formula is the actual number of atoms of each element in a molecule of a compound.

### Calculation of empirical formulae

Analysis of an organic compound may only provide information about the ratio of the numbers of atoms present. Therefore, all that can be established is its empirical formula. This might be the same as the molecular formula (as is the case with methane), but usually it is not possible to be certain.

The number of grams of each element in a sample can be determined by experiment. To find the empirical formula, the number of grams of each component must be converted into an amount in moles.

Suppose a sample of a compound contains 1.5 g of carbon and 0.5 g of hydrogen.

$$\text{amount in moles} = \frac{\text{mass in grams}}{\text{molar mass}}$$

Therefore, the amount in moles of carbon must be:

$$\frac{1.5\,\text{g}}{12\,\text{g mol}^{-1}} = 0.125\,\text{mol}$$

The amount in moles of hydrogen must be:

$$\frac{0.5\,\text{g}}{1\,\text{g mol}^{-1}} = 0.5\,\text{mol}$$

Therefore, the ratio of moles of carbon to moles of hydrogen is 0.125:0.5. This compound contains four times as many moles of hydrogen as it does carbon. Therefore, each molecule contains four times as many hydrogen atoms as it does carbon atoms.

The empirical formula of the compound is $CH_4$.

The following worked examples illustrate how experiments of this type are interpreted.

### Worked example 1

A 2.8 g sample of a hydrocarbon is found to contain 2.4 g of carbon and 0.4 g of hydrogen. What is its empirical formula?

**Answer**

$$\text{amount in moles of carbon} = \frac{2.4\,\text{g}}{12.0\,\text{g mol}^{-1}} = 0.2\,\text{mol}$$

$$\text{amount in moles of hydrogen} = \frac{0.4\,\text{g}}{1.0\,\text{g mol}^{-1}} = 0.4\,\text{mol}$$

The ratio of carbon atoms:hydrogen atoms is 0.2:0.4.
For every 1 carbon atom there are 2 hydrogen atoms.
The empirical formula is $CH_2$.
The empirical formula, $CH_2$, cannot be the molecular formula of the compound because a carbon atom has to have four bonds. Many compounds have this empirical formula — for example, it could be any member of the alkene or cycloalkane homologous series.

### Worked example 2

A sample of a compound is found to contain 3.13 g of carbon, 0.783 g of hydrogen and 2.09 g of oxygen. What is its empirical formula?

**Answer**

$$\text{amount in moles of carbon} = \frac{3.13\,\text{g}}{12.0\,\text{g mol}^{-1}} = 0.261\,\text{mol}$$

$$\text{amount in moles of hydrogen} = \frac{0.783\,\text{g}}{1.0\,\text{g mol}^{-1}} = 0.783\,\text{mol}$$

$$\text{amount in moles of oxygen} = \frac{2.09\,\text{g}}{16.0\,\text{g mol}^{-1}} = 0.131\,\text{mol}$$

The ratio of moles is not obvious. To find the simplest ratio, divide by the smallest number:

$$\text{amount of carbon} = \frac{0.261}{0.131} = 1.99 \text{ mol}$$

$$\text{amount of hydrogen} = \frac{0.783}{0.131} = 5.98 \text{ mol}$$

$$\text{amount of oxygen} = \frac{0.131}{0.131} = 1.00 \text{ mol}$$

The whole number ratio of carbon:hydrogen:oxygen is 2:6:1.
The empirical formula is $C_2H_6O$.

## Worked example 3

An oxide of nitrogen contains the following proportions by mass: 63.6% nitrogen and 36.4% oxygen. What is its empirical formula?

**Answer**

In every 100 g of the compound there is 63.6 g of nitrogen and 36.4 g of oxygen.

$$\text{amount in moles of nitrogen} = \frac{63.6\,g}{14.0\,g\,mol^{-1}} = 4.54\,mol$$

$$\text{amount in moles of oxygen} = \frac{36.4\,g}{16.0\,g\,mol^{-1}} = 2.28\,mol$$

The empirical formula is $N_2O$.

*e* Do not be put off by the use of percentages instead of grams.

## Calculation of molecular formula

The steps usually involved in the calculation of a molecular formula are as follows:

- Calculate the empirical formula from the percentage composition by mass.
- Use the empirical formula unit and the relative molecular mass to deduce the molecular formula.

## Worked example

Compound A has relative molecular mass 62. Its composition by mass is carbon 38.7%; hydrogen 9.7%; oxygen 51.6%. Calculate the empirical formula and the molecular formula of compound A.

**Answer**

| Method | Carbon | Hydrogen | Oxygen |
| --- | --- | --- | --- |
| Percentage | 38.7 | 9.7 | 51.6 |
| Divide by relative atomic mass | 38.7/12.0 = 3.23 | 9.7/1.0 = 9.7 | 51.6/16.0 = 3.23 |
| Divide by the smallest | 3.23/3.23 = 1 | 9.7/3.23 = 3 | 3.23/3.23 = 1 |

The simplest ratio of C:H:O is 1:3:1. The empirical formula is $CH_3O$.
    empirical formula mass = 12 + 3 + 16 = 31

The number of empirical units needed to make up the molecular mass is found by dividing the molecular mass by the empirical mass:

$$\frac{\text{molecular mass}}{\text{empirical mass}} = \frac{62}{31} = 2$$

Therefore, the molecular formula is made up of two empirical units of $CH_3O$. The molecular formula is $C_2H_6O_2$.

## Percentage yield

Not all reactions result in the complete conversion of reactants to products. Many preparations of organic substances are inefficient and yield only a small amount of product. Sometimes, losses occur as a result of experimental error. In these cases, the percentage yield of the reaction is recorded:

$$\% \text{ yield} = \frac{\text{amount (in moles) of product obtained}}{\text{theoretical amount (in moles) of product}} \times 100$$

Percentage yield can also be expressed as:

$$\% \text{ yield} = \frac{\text{mass of product obtained}}{\text{theoretical mass of product}} \times 100$$

Percentage yield calculations involve using amounts in moles (see page 47).

> ### Worked example 1
> Ethanol, $C_2H_5OH$, can be oxidised to form ethanoic acid, $CH_3COOH$. In an experiment, 2.3 g of ethanol is oxidised to produce 2.4 g of ethanoic acid. Calculate the percentage yield.
>
> **Answer**
> In this example, the symbol [O] is used to represent the oxidising agent.
> $$C_2H_5OH + 2[O] \rightarrow CH_3COOH + H_2O$$
> The equation shows that 1 mol of ethanol produces 1 mol of ethanoic acid.
> **Step 1:** calculate the number of moles of ethanol used
>
> $$\text{amount in moles of ethanol used} = n = \frac{\text{mass (g)}}{\text{molar mass (g mol}^{-1})}$$
>
> $$= \frac{2.3}{46} = 0.050 \, \text{mol}$$
>
> The mole ratio ethanol:ethanoic acid is 1:1, hence the maximum amount of ethanoic acid that could be made is 0.050 mol.
> **Step 2:** calculate the number of moles of ethanoic acid produced
>
> $$\text{amount in moles of ethanoic acid produced} = n = \frac{\text{mass (g)}}{\text{molar mass (g mol}^{-1})}$$
>
> $$= \frac{2.4}{60} = 0.040 \, \text{mol}$$
>
> **Step 3:** calculate the percentage yield
> $$\text{percentage yield} = \frac{\text{actual yield}}{\text{maximum yield}} \times 100 = \frac{0.040}{0.050} \times 100 = 80\%$$

*e* For mole calculations, you need to be able to write a balanced equation, so that you can use the mole ratios that it provides.

*e* It does not matter if you are unfamiliar with this reaction; simply focus on the method.

*e* $n = m/M_r$

In worked example 1 there is one organic starting material and one product, and it is assumed that there is enough oxidising agent for the reaction to go to completion. If there are two reagents, it may first be necessary to establish whether the reagents are in the exact proportions for reaction or whether one of them is in excess.

---

**Worked example 2**

Ethanol reacts with ethanoic acid according to the following equation:

$$C_2H_5OH + CH_3COOH \rightarrow CH_3COOC_2H_5 + H_2O$$

When 5.00 g of ethanol ($C_2H_5OH$) reacts with 8.00 g of ethanoic acid ($CH_3COOH$), 7.12 g of ethyl ethanoate ($CH_3COOC_2H_5$) is produced. What is the percentage yield of the reaction?

**Answer**

First, establish whether the ethanol and ethanoic acid are in the exact proportions for the reaction The equation indicates that this 1:1.

**Step 1:** calculate the number of moles of each reactant

$$\text{amount in moles of ethanol, } n = \frac{\text{mass (g)}}{\text{molar mass (g mol}^{-1})} = \frac{5.00}{46} = 0.109 \text{ mol}$$

$$\text{amount in moles of ethanoic acid, } n = \frac{\text{mass (g)}}{\text{molar mass (g mol}^{-1})} = \frac{8.00}{60} = 0.133 \text{ mol}$$

Therefore, there is an excess of ethanoic acid present. The maximum yield of the reaction is 0.109 mol of ethyl ethanoate (since 1 mol of ethanol produces 1 mol of ethyl ethanoate).

**Step 2:** calculate the number of moles of ethyl ethanoate produced

$$\text{amount of ethyl ethanoate acid produced, } n = \frac{\text{mass (g)}}{\text{molar mass (g mol}^{-1})}$$

$$= \frac{7.12}{88} = 0.0809 \text{ mol}$$

**Step 3:** calculate the percentage yield

$$\text{percentage yield} = \frac{\text{actual yield}}{\text{maximum yield}} \times 100 = \frac{0.0809}{0.109} \times 100 = 74.2\%$$

---

**Worked example 3**

In a reaction between methane and chlorine, the percentage yield of tetra-chloromethane, $CCl_4$, is 72%. Calculate the mass of tetrachloromethane that could be obtained by reacting 10.0 g of methane with excess chlorine.

**Answer**

The equation of the reaction is;

$$CH_4 + 4Cl_2 \rightarrow CCl_4 + 4HCl$$

**Step 1:** calculate the number of moles of methane present

$$\text{amount in moles of methane} = \frac{10}{16} = 0.625 \text{ mol}$$

Chlorine is in excess. Therefore, the equation indicates that if the reaction were 100% efficient, 0.625 mol of tetrachloromethane would be produced.

**Step 2:** calculate the amount in moles of tetrachloromethane produced

percentage yield = 72%

amount in moles of tetrachloromethane produced

$$= \frac{72}{100} \times 0.625 = 0.45 \, mol$$

**Step 3:** calculate the mass of tetrachloromethane produced

mass of 1 mol of $CCl_4$ = 154 g

mass of 0.45 mol = 0.45 × 154 = 69.3 g

## Atom economy

The percentage yield of a reaction is one measure of the efficiency of the process. However, a reaction with a good yield of product can still be inefficient and wasteful if there are other products that have to be discarded.

Consider the substitution reaction of sodium hydroxide and 1-bromobutane. This produces butan-1-ol and sodium bromide. The equation is:

$$C_4H_9Br + NaOH \rightarrow C_4H_9OH + NaBr$$

A percentage yield of butanol of 70% may be a good yield, but the process might be considered less than satisfactory if there were no use for the other product, sodium bromide. There is, therefore, another measure that can be applied to reactions to take account of wastage. This is called the **atom economy**.

$$atom\ economy = \frac{molecular\ mass\ of\ the\ desired\ product}{sum\ of\ the\ molecular\ masses\ of\ all\ products} \times 100$$

In the butanol example given above, the atom economy would be calculated as follows:

$M_r$ of butanol is $(4 \times 12) + (9 \times 1) + 16 + 1 = 74$

$M_r$ of sodium bromide is $23 + 80 = 103$

$$atom\ economy = \frac{74}{74 + 103} \times 100 = 42\%$$

This result indicates that the reaction is not ideal.

A reaction can be inefficient but have a high atom economy. For example, the yield of the ester, ethyl ethanoate, from the reaction between ethanol and ethanoic acid is usually only about 40%.

The equation is as follows:

$$CH_3COOH + C_2H_5OH \rightarrow CH_3COOC_2H_5 + H_2O$$

However, the atom economy is good:

$M_r$ of ethyl ethanoate = 88

$M_r$ of water = 18

$$atom\ economy = \frac{88}{88 + 18} \times 100 = 83\%$$

Addition reactions of alkenes have a 100% atom economy because there is only one product, for example:

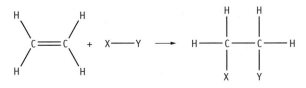

However, the percentage yield may be less than 100%.

Using industrial procedures with a high atom economy would produce less waste. Atom economy is a new concept, but its importance is likely to grow as society becomes more concerned about the need to conserve resources and avoid unwanted by-products.

## Questions

**1** Identify the functional group in each of the following molecules:
   **a** dodecane
   **b** pentene
   **c** propane
   **d** propan-2-ol
   **e** 2-bromobutane
   **f** cyclobutane

**2** For each of the following, give the formula of the next member in the homologous series:
   **a** $C_2H_5OH$
   **b** $C_3H_6$
   **c** $CH_3I$
   **d** $C_7H_{16}$

**3** Draw the displayed formula and the structural formula for the following molecules:
   **a** 2-chloropropane
   **b** 1-chloropropane
   **c** butan-2-ol
   **d** 2-methylpentane
   **e** 3-methylbut-1-ene

**4** Name the following organic compounds:
   **a**

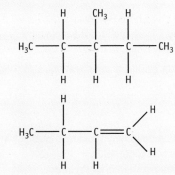

   **b**

**c**

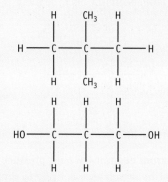

**d**

**5** Name the following compounds:
  **a** $CH_3CH_2CH_2CH_3$
  **b** $CH_3CHBrCH_3$
  **c** $CH_3CHCHCH_3$
  **d** $(CH_3)_4C$
  **e** $CH_3CH_2CH(CH_3)CH_2CH_3$

**6** Draw the skeletal formula of:
  **a** butane
  **b** but-2-ene
  **c** 2-bromopropane
  **d** propan-2-ol
  **e** propan-1-ol

**7** Draw and name the isomers of $C_5H_{12}$.

**8** Illustrating your answers by drawing appropriate isomers, explain what is meant by:
  **a** structural isomerism
  **b** *E/Z* isomerism

**9** Draw and name all the isomers of $C_4H_8$.

**10 a** Calculate the empirical formula for the following compounds:
  **(i)** Compound A contains the following by mass: carbon 37.5%; hydrogen 12.5%; oxygen 50.0%.
  **(ii)** Compound B contains the following by mass: carbon 29.3%; hydrogen 5.7%; bromine 65.0%.
  **(iii)** Compound C contains the following by mass: carbon 53.3%; hydrogen 11.1%; oxygen 35.6%.
  **b** Compound B has two isomers. Draw and name both isomers.
  **c** The relative molecular mass of compound C is 90. Deduce the molecular formula of compound C. Draw and name any isomers.

**11** 3.28 g of cyclohexene, $C_6H_{10}$, is prepared from 10.00 g of cyclohexanol, $C_6H_{11}OH$:
  $$C_6H_{11}OH \rightarrow C_6H_{10} + H_2O$$
  **a** Calculate the percentage yield.
  **b** Calculate the atom economy.

**12** The compound 2-chloro-2-methylpropane is prepared by shaking 2-methyl-propan-2-ol and concentrated hydrochloric acid together.
  **a** Write a balanced equation for the reaction between 2-methylpropan-2-ol and concentrated hydrochloric acid. Water is also produced.

**b** In the reaction, 4.0 g of 2-chloro-2-methylpropane is produced. This is equivalent to a 40% yield. Calculate the mass of 2-methylpropan-2-ol that is used.

**13** Bromobutane can be prepared by the following reaction:

$$C_4H_9OH + NaBr + H_2SO_4 \rightarrow C_4H_9Br + NaHSO_4 + H_2O$$

Calculate the atom economy for this reaction.

**14** Compounds A–F are saturated hydrocarbons.

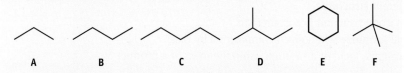

| A | B | C | D | E | F |

**a** The general formula of an alkane is $C_nH_{2n+2}$.
- **(i)** Which of the compounds A–F has a formula that does *not* fit this general formula?
- **(ii)** A, B and C are successive members of the alkane series. What is the molecular formula of the next member of the series?
- **(iii)** What is the empirical formula of compound B?

**b**
- **(i)** Three of compounds A–F are structural isomers of each other. Identify these by letter.
- **(ii)** Explain what is meant by the term 'structural isomer'.

**c** Calculate the percentage by mass of carbon in compound F.

**15** Compound X is a bromoalkene. It has the following percentage composition by mass: carbon 12.9%; hydrogen 1.1%; bromine 86.0%.

**a**
- **(i)** Calculate the empirical formula of compound X. Show your working.
- **(ii)** The relative molecular mass of compound X is 279. Show that the molecular formula is $C_3H_3Br_3$.

**b** Compound X is one of six possible structural isomers of $C_3H_3Br_3$ that are bromoalkenes. Two of these isomers are shown below as isomer 1 and isomer 2.

| Isomer 1 | Isomer 2 |

- **(i)** Draw two other structural isomers of $C_3H_3Br_3$ that are bromoalkenes.
- **(ii)** Draw two other structural isomers of $C_3H_3Br_3$ that are not bromoalkenes.
- **(iii)** Name isomer 1.

**c** 2.790 g of compound X reacts with hydrogen, $H_2$, to produce 0.843 g of 1,1,2-tribromopropane.
- **(i)** Calculate the percentage yield of 1,1,2-tribromopropane.
- **(ii)** Identify compound X by name and by formula.
- **(iii)** Calculate the atom economy of this reaction.

# Summary

You should now be able to:

- explain what is meant by the terms 'homologous series' and 'functional group'
- recall the names of the first ten alkanes
- understand how to name organic compounds
- understand the difference between empirical formulae, molecular formulae, skeletal formulae, structural formulae and displayed formulae
- understand that isomerism occurs when compounds of the same molecular formula have different structures
- explain structural isomerism and *E/Z* isomerism
- calculate empirical formulae when given composition by mass
- calculate the percentage yield of a reaction using the mass of a product
- understand the reasons for the estimation of atom economy
- calculate the atom economy of a reaction
- explain what is meant by σ- and π-bonds and the terms 'saturated' and 'unsaturated'
- understand the terms 'radical', 'electrophile' and 'nucleophile'
- understand the use of curly arrows and dipoles in discussing reaction mechanisms

# Additional reading

## 8

You may think that the IUPAC naming system for organic compounds will help you to understand the lists of contents on cosmetics and processed foods. However, although examiners use the correct systematic names in setting questions, the wider world has taken little interest, even though the IUPAC system was laid down in the 1970s.

Manufacturers have advanced beyond the first attempt at agreed naming, proposed in the late eighteenth century by Louis-Bernard Guyton de Morveau, who hoped that his system would 'help the intelligence and relieve the memory'. Although sulphuric acid is no longer referred to by the picturesque name 'oil of vitriol', manufacturers continue to use pre-1970s names. Thus, a bottle of vinegar is said to contain acetic acid, rather than ethanoic acid. Ethanoic acid is usually supplied to laboratories under the old name too.

Antifreeze for a car is referred to as glycol or ethylene glycol instead of the correct name, ethane-1,2-diol. The alcohol in drinks is just called alcohol, even though it is just one member of this homologous series. Sometimes, it is written as ethyl alcohol. This hints that the molecule contains two carbon atoms, but the correct name 'ethanol' should be used. As well as ethanol, wine contains acids, such as tartaric acid, more correctly known as 2,3-dihydroxybutanoic acid. Small crystals may form at the bottom of a wine bottle, owing to precipitation of potassium hydrogentartrate, which is known to cooks as 'cream of tartar'.

Sometimes the systematic name is too long for general usage. For example, to say a person has a raised 10,13-dimethyl-17-(6-methylheptan-2-yl)-2,3,4,7,8,9,11, 12,14,15,16,17-dodecahydro-1*H*-cyclopenta[*a*]phenan-thren-3-ol level is more cumbersome than 'raised cholesterol'. This can apply to the E-numbers used in foods, which vary from the simple (E290, carbon dioxide; E520, magnesium oxide) to the complex (E612, the flavour enhancer monosodium glutamate — the monosodium salt of 2-aminopentanedioic acid). Biologists deal with complex substances and use the old names — for example, 'glycine', rather than amino-ethanoic acid for the simplest amino acid.

The old names have influenced the newer system. For example, we refer to poly(ethene) or polythene, which indicates that it has been made by combining many ethene molecules. A better name would be polythane because the polymer does not contain any double bonds

Potential confusion is not restricted to organic compounds. Some cookery books list sodium bicarbonate as an ingredient. This suggests that it contains two carbonate groups for each sodium atom — a problem solved by using the correct name, sodium hydrogencarbonate.

## Extension question: isomers

It is difficult to be sure that you have identified all the isomers for a given formula. It also easy to write two formulae that are, in fact, the same.

a Draw all the isomers of $C_5H_{10}$ that are alkenes. Identify any that are *E/Z* isomers.

b Name the isomers in part a.

c Draw all the isomers of formula $C_2ClBrIF$.

Naming the compounds in part c requires the following rules. For each carbon atom on either side of the double bond, determine the atomic numbers of the substituents. If the two highest numbers are on the same side of the molecule, the compound is a (*Z*)-isomer; if they are on opposite sides, it is an (*E*)-isomer.

For example, consider the compound:

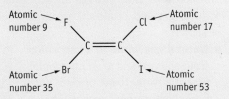

Here, the two atoms with the highest atomic numbers, Br and I, are on the same side of the molecule — the molecule is named '*Z*'. If they had been on different sides, it would have been named '*E*'.

The full name is (*Z*)-1-bromo-2-chloro-1-fluoro-2-iodoethene. Substituents are listed in alphabetical order and the relevant numbers are then added to show their positions.

d Using the above example as a guide, name the isomers you have drawn in answer to part c.

# Fuels, alkanes and alkenes

## Chapter 9

The Earth has large, though depleting, reserves of hydrocarbons. These compounds are particularly important as fuels. Hydrocarbons are compounds that contain only hydrogen and carbon. The carbon atoms in hydrocarbons can be linked by single, double or triple bonds and can be in a chain or a ring. This chapter examines the production of hydrocarbon fuels and considers the alternatives that could replace them in the future. The properties of two classes of hydrocarbon — alkanes and alkenes — are also considered.

A fuel is a source of energy. A useful fuel should:

■ have a high-energy output
■ be either abundant or manufactured easily
■ be easy to transport and store
■ be easy to ignite
■ cause minimum environmental damage

## Hydrocarbons as fuels

A **fossil fuel** is a non-renewable energy source, which is the product of the sun's energy used by plants for photosynthesis, millions of years ago. The fuel is produced over time through the anaerobic decay of plant and animal material. Natural gas and crude oil are fossil fuels that have been the main source of alkanes:

■ **Natural gas** varies in composition from one gas field to another, but is essentially methane mixed with other hydrocarbons such as ethane, propane and butane. It usually contains some hydrogen sulphide. Natural gas is processed and the methane is separated before it is supplied to homes.
■ **Crude oil** is a complex mixture of hydrocarbons, the composition of which varies. North Sea oil, for example, is richer in gasoline and naphtha than oils found in the Americas.

*An off-shore oil extraction platform*

# Refining crude oil

Crude oil is refined so that the different hydrocarbons in the mixture can be separated. The initial separation is carried out by fractional distillation.

## Fractional distillation of crude oil

**Fractional distillation** is a method used to separate a mixture of liquids that have different boiling points. The fractional distillation of crude oil is a continuous process that separates the crude oil into fractions, each of which contains a mixture of hydrocarbons within a narrow range of boiling points. In general, components that contain more carbon atoms boil at a higher temperature because the intermolecular forces (usually van der Waals forces) are stronger.

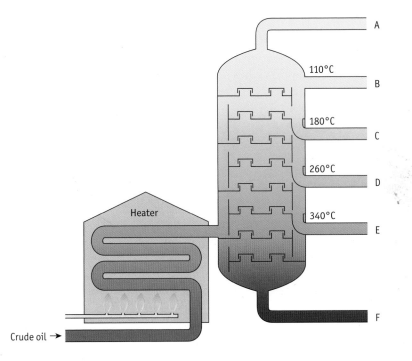

*Figure 9.1 Fractional distillation of crude oil*

The fractions produced are summarised in Table 9.1. The letters A–F refer to Figure 9.1.

|   | Length of carbon chain | Fraction | Use |
|---|---|---|---|
| A | 1–4 | Refinery gas | Natural gas; LPG |
| B | 5–10 | Gasoline | In petrol (cars) |
| C | 8–10 | Naphtha | Cracking, feedstock |
| D | 11–16 | Kerosine | Jet fuel; heating oil |
| E | 13–25 | Diesel | Diesel (cars and lorries); heating oil |
| F | > 25 | Residue | Fuel oil for power stations; waxes; bitumen |

*Table 9.1 Fractions from crude oil*

LPG stands for liquefied petroleum gas. This is a mixture of propane and butane.

These fractions can be used directly or refined further for use as fuels or as feedstocks for the petrochemical industry.

■ The gasoline fraction (alkanes containing between five and ten carbon atoms) are used as petrol.
■ The naphtha fraction (alkanes that contain between eight and ten carbon atoms) is most important as a source of chemicals for the chemical industry.

The gasoline and naphtha fractions obtained from the fractional distillation of crude oil are insufficient to meet demand. Some of the heavier, less volatile fractions are treated (cracked) to generate more of the high-demand gasoline and naphtha.

## Cracking

Cracking involves heating the heavier oil fractions in the presence of a catalyst, so that long-chain molecules are broken down into shorter and more useful molecules. The catalyst used can be either a mixture of silicon and aluminium oxides or a zeolite (see page 148). The process involves the random breaking of C–C and C–H bonds and can lead to a wide variety of products. It can be summarised as:

long chain → shorter chain + alkene
alkane            alkane

The resulting alkenes are used as feedstocks in the chemical industry to make a variety of products.

The breaking of bonds in a long-chain alkane is random. The long-chain alkane undecane ($C_{11}H_{24}$) can be cracked to yield different products:

*Some products made from alkenes*

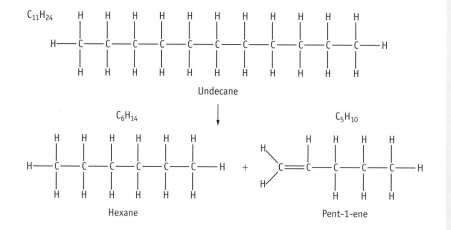

A second way that undecane could be cracked is:

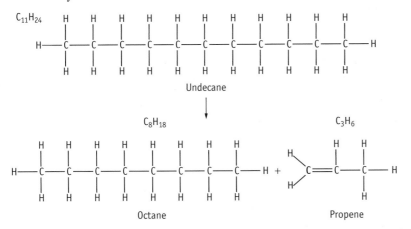

A third way that undecane could be cracked is:

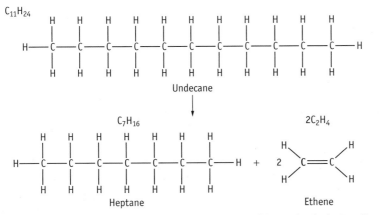

Modern car engines require higher proportions of branched-chain alkanes, cycloalkanes and arenes, which burn more smoothly than straight-chain alkanes. Straight-chain alkanes often pre-ignite, which leads to 'knocking' or 'pinking', so they are processed into branched-chain alkanes, cycloalkanes or arenes.

Isomerisation

Isomerisation involves changing an alkane into an isomer. Straight-chain alkanes are converted into branched-chain alkanes by passing their vapour over a hot platinum catalyst:

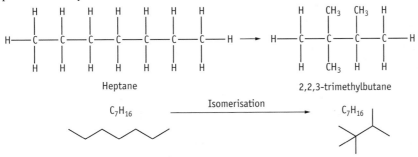

The resulting mixture contains straight-chain and branched-chain alkanes, which are separated using a molecular sieve. The sieve is a zeolite — an aluminosilicate with a regular network of pores (or holes). The size of the holes is critical because the straight-chain alkanes pass through the zeolite but the branched-chain isomers are too bulky, allowing them to be separated.

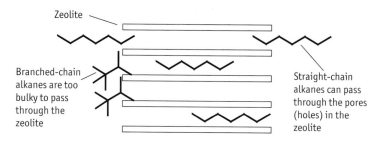

Zeolite

Branched-chain alkanes are too bulky to pass through the zeolite

Straight-chain alkanes can pass through the pores (holes) in the zeolite

Zeolites are complex structures of aluminium, silicon and oxygen atoms, with charge-balancing metal cations. Some occur naturally in volcanic rock and ash. Synthetic zeolites are used as catalysts for chemical reactions that take place within their internal cavities. The structure of a zeolite means that only certain molecules can pass into its cavities.

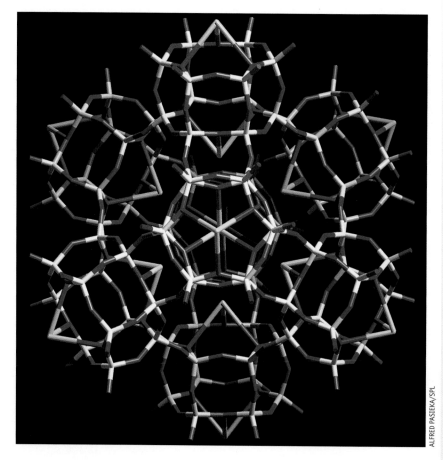

*Structure of a zeolite*

ALFRED PASIEKA/SPL

## Reforming

Reforming is a process used to obtain cycloalkanes and arenes from straight-chain alkanes. Reforming reactions are catalysed by bimetallic catalysts. The bimetallic catalyst usually consists of two transition elements, such as platinum and rhenium or platinum and iridium.

The reforming of hexane into cyclohexane is shown below:

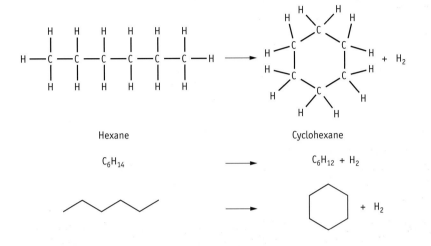

Hexane            Cyclohexane

$C_6H_{14}$            $C_6H_{12} + H_2$

Reforming always produces hydrogen as a co-product. The reforming of hexane, $C_6H_{14}$, can eventually produce benzene, $C_6H_6$. This process is represented in skeletal form below:

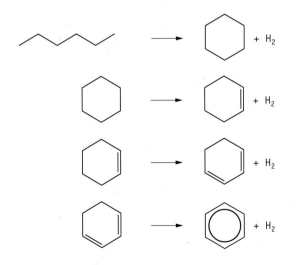

The overall equation can be represented as:

Summary of the refining of crude oil

**Cracking** — long chain alkenes are broken down into more useful shorter-chain alkanes and alkenes. The shorter-chain alkanes are used in petrol. The alkenes are used to make a wide range of useful products such as alcohols and polymers.

**Isomerisation** — straight-chain alkanes are turned into branched-chain alkanes, which burn more smoothly and are, therefore, better and more efficient fuels.

**Reforming** — straight-chain alkanes are turned into cycloalkanes or arenes that also burn more smoothly and are, therefore, better and more efficient fuels. Hydrogen is always produced as a co-product. Branched-chain compounds, cycloalkanes and arenes are added to petrol to increase its octane rating.

# Alternative fuels

Fossil fuels, such as gas and oil, are essential to society and are used extensively in industry, in the home and in transport. However, fossil-fuel reserves are limited and will eventually run out because they are non-renewable. When fossil fuels are exhausted, alternative sources of energy will be needed by society. In addition, the combustion of fossil fuels generates carbon dioxide gas, which is a greenhouse gas believed to be the key contributor to global warming.

Chemists and other scientists have been working for many years to develop alternative fuels. Currently, research is being carried out into ways of harnessing the energy in natural sources such as solar, tidal and wind energy. Additionally, much effort is being put into the development of biofuels, nuclear fission and nuclear fusion.

## Biofuels

**Biofuels**, such as bioethanol and biodiesel, are made from crops — for example wheat and sugar beet — and from vegetable oils. Bioethanol and biodiesel are environmentally friendly fuels because they are classed as 'carbon-neutral'. Burning biofuels simply sends back into the atmosphere the carbon dioxide that the plants took out when they were growing in the fields. According to the UK government agency, the Central Science Laboratory, bioethanol made from grain produces 65% less greenhouse gas than petrol. However, if biofuels are to replace fossil fuels, car engines will have to be adapted and more arable land will be needed to grow the source crops.

Bioethanol

**Bioethanol** is produced from biomass by hydrolysis and sugar fermentation processes. The main sources of biomass are maize, wheat and waste straw. There is also research into the use of municipal solid wastes for ethanol production. Up to 5% bioethanol can be mixed with petrol for use in cars that run on ordinary unleaded petrol. Bioethanol mixed with petrol is now sold at a number of filling stations in the UK.

*Saab is pioneering the use of bioethanol as an eco-friendly, renewable energy source*

*Bioethanol plant, with a field of maize in the foreground*

### Biobutanol

**Biobutanol** is another alcohol-based fuel that can be mixed with petrol. It is made by fermentation using bacterial enzymes. It is produced from the same agricultural feedstocks as bioethanol, but it can be blended with petrol at higher concentrations. Up to 10% biobutanol can be incorporated into ordinary petrol. This can be used in any car that runs on unleaded petrol.

## Biodiesel

**Biodiesel** is produced from vegetable crops such as oilseed rape, or by the conversion of waste materials such as cooking oil. Up to 5% biodiesel can be mixed with conventional diesel. It is already sold in the UK under the brand name GlobalDiesel.

*Oilseed rape is one of a range of crops that are suitable for the production of biodiesel*

## Air travel: the future

In 2002, the UK's Royal Commission on Environmental Pollution predicted that by 2050, air travel will account for 75% of the UK's greenhouse carbon emission. Commercial jets use a petroleum fuel called Jet A, which remains liquid at extreme temperatures (down to −40°C). Biofuel alternatives, made from vegetable oil, freeze at a much higher temperature and research is being carried out to try to solve this problem.

## Questions

1 Dodecane, $C_{12}H_{26}$, can be cracked in a number of ways. Write balanced equations for each of the following:
   a dodecane cracked to produce nonane and propene only
   b dodecane cracked to produce octane and ethene only
   c dodecane cracked to produce heptane, propene and ethene only

2 Hexane can be isomerised into branched-chain alkanes. How many different isomers could be produced?

3 Heptane can be isomerised/reformed into cyclic compounds. How many different compounds with a six-membered ring could be produced? Write a balanced equation for each reaction.

4 Petrol is a complex mixture of more than 100 different compounds. Most of the compounds come from the gasoline fraction obtained from the primary fractional distillation of crude oil. However, the gasoline fraction accounts for only about 25% of crude oil. Oil companies have found ways to:
   ■ use the remaining fractions from the crude oil
   ■ refine these compounds into the blend needed to produce an efficient fuel

The co-products of some of these processes are used to make a number of useful substances. Some of these processes are illustrated below:

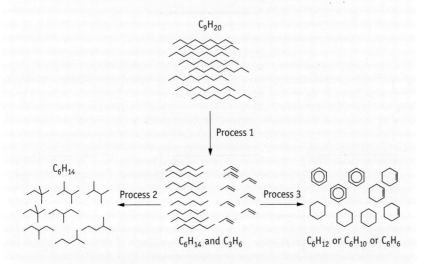

Name, describe and state a use for each of the three processes shown above. Include relevant balanced equations for each process in your answer.

# Alkanes

**Alkanes** are called **saturated** hydrocarbons because all the C–C bonds are single bonds.

## Physical properties

### Melting point and boiling point

The variation in boiling point depends on the degree of intermolecular bonding (page 91). The only intermolecular bonding that occurs in alkanes are van der Waals forces. There are two important trends in the variation of boiling point:

- As the relative molecular mass of the alkanes increases, so does the boiling point. This is due to an increase in chain length and an increase in the number of electrons, both of which strengthen the van der Waals forces. The boiling points of the first six alkanes are shown in Table 9.2 and the trend is shown graphically in Figure 9.2.

| Alkane | Molecular formula | Boiling point/K | Boiling point/°C | State at r.t.p. |
|---|---|---|---|---|
| Methane | $CH_4$ | 109 | −164 | Gas |
| Ethane | $C_2H_6$ | 185 | −88 | Gas |
| Propane | $C_3H_8$ | 231 | −42 | Gas |
| Butane | $C_4H_{10}$ | 273 | 0 | Gas |
| Pentane | $C_5H_{12}$ | 309 | 36 | Liquid |
| Hexane | $C_6H_{14}$ | 342 | 69 | Liquid |

*Table 9.2 Boiling points of the alkanes*

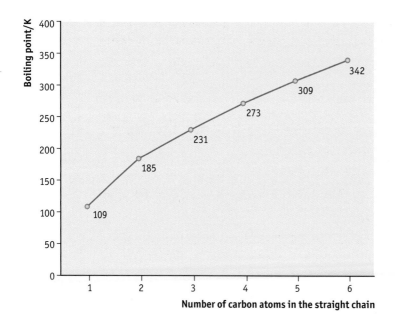

*Figure 9.2* Trend
in boiling point of
the first six alkanes

- For isomers with the same relative molecular mass, the boiling point decreases as the degree of branching increases. This is illustrated in Table 9.3 for the three isomers of $C_5H_{12}$, which have relative molecular mass 72.

| Name | Structure | Boiling point/K | Boiling point/°C | State at r.t.p. |
|------|-----------|-----------------|------------------|-----------------|
| Pentane | | 309 | 36 | Liquid |
| Methylbutane | | 301 | 28 | Liquid/gas |
| Dimethylpropane | | 283 | 10 | Gas |

*Table 9.3*
*Relationship*
*between boiling*
*point and degree*
*of branching*

As shown in the diagram below, molecules of straight-chain alkanes, such as pentane, pack together more closely than molecules of branched-chain alkanes, such as methylbutane and dimethylpropane. This increased amount of surface contact creates more intermolecular forces. More energy is needed to overcome these intermolecular forces, so the boiling point of pentane is the highest of the three isomers.

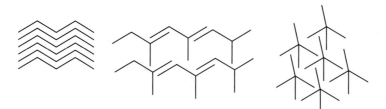

Pentane                    Methylbutane                    Dimethylpropane

Branched-chain alkanes, such as methylbutane and dimethylpropane, cannot pack together as tightly, reducing the intermolecular forces and, hence, the boiling point and melting point.

### Solubility
Alkanes are non-polar. They cannot form links with molecules that are polar, which means that they do not mix with polar molecules. Therefore, alkanes are insoluble in polar solvents, such as water. Liquid alkanes are less dense than water and form an immiscible upper layer.

Alkanes dissolve readily in each other because the molecules link through van der Waals forces. Crude oil is a mixture, or solution, of over 100 different alkanes.

## Chemical properties

Since they are non-polar, alkanes are relatively unreactive. However, they do undergo two important reactions — combustion and radical substitution.

### Combustion of alkanes
Combustion of alkanes in excess oxygen produces carbon dioxide and water:

$$CH_4 + 2O_2 \rightarrow CO_2 + 2H_2O$$
$$C_2H_6 + 3\tfrac{1}{2}O_2 \rightarrow 2CO_2 + 3H_2O$$
$$C_{11}H_{24} + 17O_2 \rightarrow 11CO_2 + 12H_2O$$

Combustion of alkanes in a limited supply of oxygen produces carbon monoxide and water. Carbon monoxide is poisonous and restricts the effectiveness of the blood to transport oxygen in the body. Therefore, when alkanes are burnt there must be good ventilation, and car engines must be effectively tuned to ensure that the volume of carbon monoxide in the exhaust is within prescribed limits.

$$CH_4 + 1\tfrac{1}{2}O_2 \rightarrow CO + 2H_2O$$
$$C_2H_6 + 2\tfrac{1}{2}O_2 \rightarrow 2CO + 3H_2O$$
$$C_{11}H_{24} + 11\tfrac{1}{2}O_2 \rightarrow 11CO + 12H_2O$$

### Radical substitution
A radical is a particle with an unpaired electron. The unpaired electron in a free radical is usually shown as a dot, e.g. $Cl\bullet$, $\bullet CH_3$. Radicals are highly reactive and can attack non-polar compounds such as alkanes. The mechanism of such reactions (i.e. the steps involved) is important and illustrates some general features of radical chemistry.

#### *Substitution by chlorine or bromine to form halogenoalkanes*
The reaction between methane and chlorine is used as an example.

**Reagent:** chlorine
**Conditions:** ultraviolet radiation
**Equation:** $CH_4 + Cl_2 \rightarrow CH_3Cl + HCl$
**Mechanism:** radical substitution

There are three distinct steps that make up this mechanism:
- initiation
- propagation
- termination

In the **initiation** step, free radicals are generated. The ultraviolet radiation provides sufficient energy to break the Cl–Cl bond homolytically to produce radicals. **Homolysis** or **homolytic fission** occurs when a covalent bond breaks and the atoms joined by the bond separate, each atom taking one of the shared pair of electrons:

$$Cl-Cl \rightarrow Cl\bullet + Cl\bullet$$

Half-arrows can be used to indicate the movement of a single electron:

$Cl-Cl \xrightarrow{\text{UV}} 2Cl\bullet$

**Propagation** involves more than one step, each of which maintains the radical concentration. Usually a $Cl\bullet$ or a $Br\bullet$ free radical causes the release of an alkyl radical or vice versa. The propagation steps are difficult to control and can lead to a complex mixture of products.

For mono-substitution there are two essential propagation steps:

**Propagation step 1:** $CH_4 + Cl\bullet \rightarrow HCl + CH_3\bullet$
**Propagation step 2:** $CH_3\bullet + Cl_2 \rightarrow CH_3Cl + Cl\bullet$

An important feature of the propagation steps is that the second reaction generates another $Cl\bullet$. This means that, in principle, the reaction might never stop and many molecules of methane might be destroyed from a single initiation. In practice, this does not occur because termination steps take place.

Termination involves any two radicals reacting together to form a covalent bond:

**Chain termination steps:** $CH_3\bullet + CH_3\bullet \rightarrow C_2H_6$
$CH_3\bullet + Cl\bullet \rightarrow CH_3Cl$
$Cl\bullet + Cl\bullet \rightarrow Cl_2$

Radicals are extremely reactive. They react with almost any molecule, atom or radical with which they collide. In the second propagation step a chlorine radical could collide with the product molecule, $CH_3Cl$, in which case the following sequence would occur:

$CH_3Cl + Cl\bullet \rightarrow HCl + \bullet CH_2Cl$
$\bullet CH_2Cl + Cl_2 \rightarrow CH_2Cl_2 + Cl\bullet$

It follows that $CH_2Cl_2$ might also react with another chlorine radical and so on. The result is that when methane reacts with chlorine, a mixture of $CH_3Cl$, $CH_2Cl_2$, $CHCl_3$ and $CCl_4$ is formed.

The overall reaction is an example of **homolytic radical substitution**. This reaction occurs with all alkanes. The sequence for ethane is:

**Initiation step:** $Cl_2 \rightarrow 2Cl\bullet$
**Propagation step 1:** $C_2H_6 + Cl\bullet \rightarrow HCl + C_2H_5\bullet$
**Propagation step 2:** $C_2H_5\bullet + Cl_2 \rightarrow C_2H_5Cl + Cl\bullet$

In this step, HCl is always formed.

In this step, $Cl\bullet$ is always formed, along with a product molecule.

The final stage involves any two radicals reacting together.

**Chain termination steps:** $C_2H_5\bullet + C_2H_5\bullet \rightarrow C_4H_{10}$
$C_2H_5\bullet + Cl\bullet \rightarrow C_2H_5Cl$
$Cl\bullet + Cl\bullet \rightarrow Cl_2$

**1 a** Draw a dot-and-cross diagram to represent a molecule of:
    **(i)** methane          **(ii)** ethane

  **b** Give the bond angles in:
    **(i)** methane          **(ii)** ethane

**2 a** What is meant by electronegativity?
  **b** Explain why alkanes are unreactive.

**3** Draw and name the isomers of $C_6H_{14}$. State which isomer has the highest boiling point and which isomer has the lowest boiling point. Explain your answers.

**4** Write balanced equations for the complete combustion of the following:
  **a** propane
  **b** pentane
  **c** cyclopentane

**5** Explain each of the following terms:
  **a** free radical
  **b** homolytic fission
  **c** initiation
  **d** propagation
  **e** termination
  **f** substitution

**6** Write the balanced equation and the mechanism for the reaction between:
  **a** ethane and bromine
  **b** cyclobutane and chlorine

**7** Calculate the formulae of alkanes A–C.
  **a** When $100\,cm^3$ of alkane A is burned in excess oxygen, $300\,cm^3$ carbon dioxide gas and $400\,cm^3$ water (steam) are produced. Calculate the formula of alkane A.
  **b** When burned, $0.1\,mol$ of alkane B produces $10.8\,g$ of water. Calculate the formula of alkane B.
  **c** When $60\,cm^3$ of alkane C reacts with $700\,cm^3$ of oxygen (an excess) and is cooled to room temperature, a gaseous mixture with a volume of $460\,cm^3$ is produced. When this gaseous mixture is passed through limewater, the volume is reduced to $40\,cm^3$.
    **(i)** Deduce the volume of carbon dioxide gas in the mixture.
    **(ii)** Deduce the volume of oxygen gas in the mixture.
    **(iii)** Calculate the formula of alkane C. Write a balanced equation for the reaction.

# Alkenes

**Unsaturated** hydrocarbons contain multiple carbon–carbon bonds. **Alkenes** are unsaturated molecules that contain C=C double bonds. They do not occur naturally and are manufactured by cracking the gasoline and naphtha fractions of crude oil. The propane and butane mixture, LPG, can also be cracked to make ethene.

# Physical properties

Boiling point

As with alkanes, the boiling points of the alkenes increase with increasing chain length. However, the boiling point of an alkene is lower than that of the corresponding alkane (Table 9.4).

| Alkane | Boiling point/K | | Alkene | Boiling point/K |
|--------|-----------------|---|--------|-----------------|
| Ethane | 185 | | Ethene | 169 |
| Propane | 231 | | Propene | 226 |
| Butane | 273 | | But-1-ene | 267 |
| Pentane | 309 | | Pent-1-ene | 303 |

*Table 9.4*
*Comparison of the boiling points of alkanes and alkenes*

The alkenes have lower boiling points because their molecules cannot pack together as efficiently as those of alkanes. Hence there are fewer intermolecular forces to overcome.

Solubility

Alkenes, like alkanes, are insoluble in water.

Bonding in alkenes

An alkene molecule contains a C=C double bond. The double bond consists of a σ-bond and a π-bond.

- A **σ-bond** is a single covalent bond made up of two shared electrons with the electron density concentrated between the two nuclei.
- A **π-bond** is formed by the 'sideways' overlap of two *p*-orbitals on adjacent carbon atoms, each providing one electron.

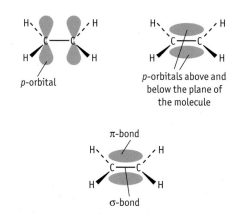

*p*-orbital

*p*-orbitals above and below the plane of the molecule

π-bond

σ-bond

The C–H bond angle at each side of the C=C double bond is approximately 120° (usually in the region 116–124°), which results in a trigonal planar arrangement of bonds round each carbon atom:

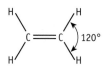

120°

In other alkenes, for example propene and but-1-ene, the equivalent bond angles on either side of the C=C double bond are approximately 120°, but the bond angles in the alkyl groups are 109.5°, as expected:

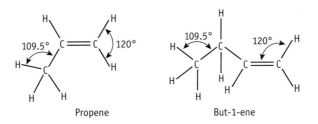

Propene                But-1-ene

The C=C double bond prevents freedom of rotation, which under certain circumstances can lead to the existence of stereoisomers, known as *E/Z* isomers.

The key features essential for *E/Z* isomers are:
- the restricted rotation due to the C=C double bond
- each carbon atom in the C=C double bond is bonded to two different atoms or groups

Details of *E/Z* isomers are given on page 130. The simplest example of *E/Z* isomerism occurs in but-2-ene. Although the difference in geometry changes the shape of the molecule, it has little or no effect on the general chemistry of the two isomers. However, (*Z*)-but-2-ene has a slightly higher boiling point than (*E*)-but-2-ene.

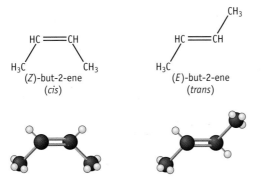

(*Z*)-but-2-ene        (*E*)-but-2-ene
(*cis*)             (*trans*)

## Chemical properties

The C=C double bond is unsaturated, which means that the main reactions of alkenes are **addition reactions** (strictly, **electrophilic addition reactions**). In this type of reaction, the π-bond breaks and an atom, or group, adds to each of the two carbon atoms. The general reaction can be summarised as:

$$CH_2=CH_2 + X–Y \rightarrow CH_2XCH_2Y$$

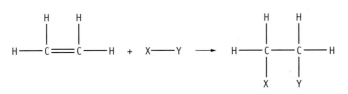

**e** For most organic reactions, you need to know:
- the balanced equation
- the reagents
- the conditions
- what is observed

Reactions of ethene

## Hydrogenation

Ethene reacts with hydrogen to form ethane, according to the following equation:

$$CH_2CH_2 + H_2 \rightarrow CH_3CH_3$$

**Reagent:** hydrogen
**Conditions:** nickel catalyst at 150°C
**Observations:** no visible change

Hydrogenation of polyunsaturated compounds derived from plant oils is used in the production of margarine (page 192).

## Bromination

Ethene reacts with bromine to produce 1,2-dibromoethane, according to the equation:

$$CH_2CH_2 + Br_2 \rightarrow CH_2BrCH_2Br$$

**Reagent:** bromine
**Conditions:** mix at room temperature
**Observations:** decolorisation of bromine

This reaction is used to test for the presence of a C=C double bond.

## Reaction of ethene with hydrogen chloride

Ethene reacts with hydrogen chloride to produce chloroethane, according to the following equation:

$$CH_2CH_2 + HCl \rightarrow CH_3CH_2Cl$$

**Reagent:** hydrogen chloride
**Conditions:** mix gases at room temperature
**Observations:** no visible change

## Hydration of ethene

A **hydration** reaction involves the addition of water and results in a new organic compound. Ethene is hydrated according to the following equation:

$$CH_2CH_2 + H_2O \rightarrow CH_3CH_2OH$$

**Reagent:** water (steam)
**Conditions:** an acid catalyst, such as phosphoric(V) acid, $H_3PO_4$, at 300°C and 6 MPa
**Observations:** no visible change

All alkenes undergo similar reactions under similar conditions. However, unsymmetrical alkenes, such as propene, produce two isomers when reacted with hydrogen chloride or water:

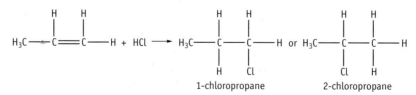

1-chloropropane          2-chloropropane

> **e** Remember that the bromine is *decolorised*. In an exam, you will be penalised if you write that it 'goes clear' or is 'discoloured'.

> **e** It is easy to confuse hydration with hydrolysis, which also involves water. However, in hydrolysis reactions, water reacts with an organic compound and *two* new compounds are produced.

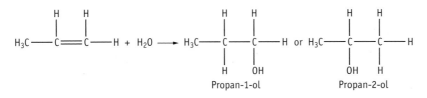

*Propan-1-ol* ... *Propan-2-ol*

### Mechanism of electrophilic addition

The mechanism of an electrophilic addition reaction involves an **electrophile** (electron-pair acceptor) and the movement of electron pairs, indicated by the use of curly arrows. The curly arrow always points from an area that is electron-rich to an area that is electron-deficient. The reaction is initiated by the presence of a dipole (charge separation along a bond) on the molecule that reacts with the alkene. This dipole could be:

- permanent, as in hydrogen chloride
- induced by the presence of the alkene, as is the case with bromine

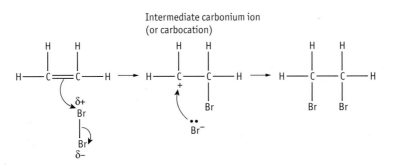

Intermediate carbonium ion (or carbocation)

The intermediate carbonium ion can also be drawn as:

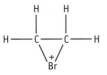

The key features of this mechanism are as follows:

- When the Br–Br molecule approaches the ethene, a temporary induced dipole is formed, resulting in polarisation of the Br–Br bond:

$$\overset{\delta+}{Br} \text{———} \overset{\delta-}{Br}$$

- The initial curly arrow starts at the $\pi$-bond (within the C=C double bond) and points to the $Br^{\delta+}$.
- The second curly arrow shows the movement of the bonded pair of electrons in the Br–Br bond towards the $Br^{\delta-}$, resulting in heterolytic fission of the Br–Br bond.
- An intermediate carbonium ion (also called a carbocation) is formed together with a $Br^-$ ion, which contains the pair of electrons from the Br–Br bond.
- The third curly arrow from the lone pair of electrons on the $Br^-$ to the positively charged carbonium ion shows the formation of 1,2-dibromoethane.

**e** When describing mechanisms, it is essential that you show:
- relevant dipoles
- lone pairs
- curly arrows

When bromine reacts with an alkene, the Br–Br bond undergoes *heterolytic* fission; when it reacts with an alkane it undergoes *homolytic* fission. Compare the two ways in which the Br–Br bond is broken:

Br —⌒— Br  $\longrightarrow$  Br$^+$ + Br$^-$     Br —⌒⌒— Br  $\longrightarrow$  Br• + Br•
Heterolytic fission                              Homolytic fission

## Addition polymerisation of alkenes

Alkenes can undergo addition reactions in which one alkene molecule joins to a another and so on until a long molecular chain is built up. The individual alkene molecule is referred to as a **monomer**; the long-chain molecule is called a **polymer.**

Polymerisation can be initiated in a number of ways. Often, the initiator is incorporated at the start of the long molecular chain. However, if the initiator is disregarded, the empirical formulae of the monomer and the polymer are the same.

### Poly(ethene)

The simplest monomer is ethene, which can be polymerised to produce poly(ethene). In the formula for poly(ethene), $n$ is a large number — it may be as high as 10 000.

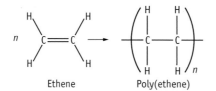

Ethene          Poly(ethene)

Poly(ethene), or polythene as it is commonly called, was developed in the 1930s. Today, it is widely used in everyday applications. It is tough, durable and unreactive. Poly(ethene) is non-polar and the molecules contain only strong covalent bonds. It is an excellent insulator and, because it is unaffected by the weather, is used for insulating electricity cables.

Changing the monomer gives polymers with different properties for different applications.

The properties of a polymer can be altered by the addition of a **plasticiser**. This is usually a liquid with a high boiling point (e.g. an ester). The plasticiser molecules are located between the polymer molecules, which reduces the intermolecular forces and makes the polymer more flexible (less rigid).

### Poly(propene)

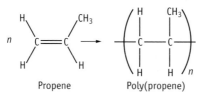

Propene          Poly(propene)

Poly(propene) is strong and durable. It is used for food boxes, clothing, ropes and carpets.

### Poly(chloroethene)

The old name for chloroethene is vinyl chloride. Hence, the product of this addition polymerisation reaction is often referred to as polyvinyl chloride, or PVC.

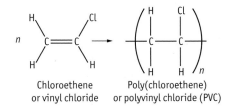

Chloroethene
or vinyl chloride

Poly(chloroethene)
or polyvinyl chloride (PVC)

Unplasticised PVC is rigid. It is used in water pipes, credit cards and window frames. Plasticised PVC is flexible and is used in raincoats, shower curtains and packaging film.

### Poly(phenylethene)

The common name for phenylethene is styrene. Therefore, the product of this polymerisation is polystyrene. The correct chemical name is poly(phenylethene).

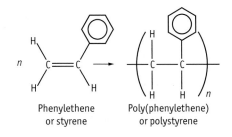

Phenylethene
or styrene

Poly(phenylethene)
or polystyrene

Polystyrene is an excellent insulator. It is resistant to acids and alkalis and does not absorb water. It is used in Styrofoam cups, fast-food containers, refrigerator insulation, packaging, telephones and plastic flowerpots.

### Poly(tetrafluoroethene), PTFE

Poly(tetrafluoroethene) is also known as Teflon®.

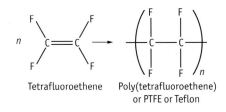

Tetrafluoroethene

Poly(tetrafluoroethene)
or PTFE or Teflon

PTFE is strong, unreactive and non-flammable and has a slippery surface. It is used for gaskets, plumbing tape and lining non-stick cookware.

### Identification of the monomer from which a polymer is produced

It is possible to deduce the repeat unit of an addition polymer and, therefore, to identify the monomer from which the polymer is produced. For example:

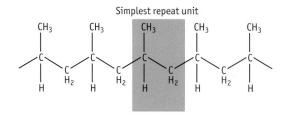

Simplest repeat unit

Therefore, the monomer is propene:

### Disposal of polymers

The widespread use of these polymers has created a major disposal problem. The bonds in addition polymers are strong, covalent and non-polar, making most of the polymers resistant to chemical attack. In addition, since they are not broken down by bacteria, they are referred to as being **non-biodegradable**.

Plastic waste is usually buried in landfill sites where it remains unchanged for decades. This means that local authorities have to find more and more locations for landfill.

An alternative to dumping is incineration. Polymers are made from hydrocarbons and are, therefore, potentially good fuels. When burned, they release useful energy. In the UK, only about 10% of plastic is incinerated, but in countries such as Japan and Denmark up to 70% is burned. Some plastics, such as PVC, produce poisonous gases (e.g. HCl), so the incinerators have to be fitted with gas-scrubbers to prevent these toxins from being released into the environment.

Another possibility is recycling the polymers and using them as feedstock for the production of new polymers. Different types of polymer have to be separated from each other because mixtures yield inferior plastic products.

Considerable effort has been put into developing biodegradable polymers that have useful properties. The aim is to create a polymer that contains an active functional group that can be attacked by bacteria. Promising candidates are based on the polymerisation of isoprene — the monomer from which rubber is made. The correct name for isoprene is 2-methylbuta-1,3-diene:

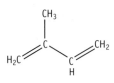

Other biodegradable compounds are based on condensation polymers, which are studied at A2.

## Questions

**1 a** Name the following alkenes:

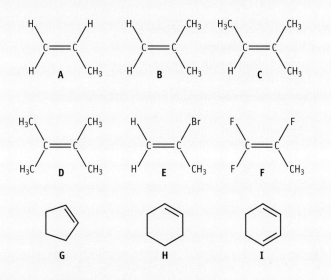

**b** What is the empirical formula of compounds A–C?
**c** What is the molecular formula of compound D?
**d** What is the molecular formula of compound I?
**e** What is the empirical formula of all alkenes?

**2 a** Write a balanced equation for the following reactions:
   **(i)** the complete combustion of propene
   **(ii)** the reaction between but-1-ene and hydrogen
   **(iii)** the hydration of but-2-ene
   **(iv)** the reaction between buta-1,3-diene and bromine
**b** State what you would observe in the reaction between buta-1,3-diene and bromine and name the organic product.
**c** Explain why the hydration of but-2-ene gives only one organic product, whereas the hydration of but-1-ene gives two organic products.

**3** Explain what is meant by the term 'electrophile'.

**4** Write a balanced equation and the mechanism for the reaction between bromine and cyclohexene. Use curly arrows to show the movement of electrons. Show any relevant dipoles and lone pairs of electrons.

**5** Alkenes can undergo polymerisation. Write an equation to show the polymerisation of but-1-ene. Draw two repeat units of the polymer.

**6 a** Explain why the disposal of polymers made from alkenes is problematic.
**b** Explain why polymers made from alkenes are good insulators.

**7** Explain how waste polymers are likely to be treated in the future in order to reduce damage to the environment.

# Summary

You should now be able to:

- understand and explain the importance of the processing of crude oil through fractional distillation, cracking, isomerisation and reforming
- understand the importance of developing alternative fuels based on renewable resources
- explain the boiling points and solubility of the hydrocarbons
- describe the combustion of alkanes and alkenes
- describe the radical substitution of alkanes with chlorine or bromine
- describe the bonding and shapes of alkenes
- describe the reactions of hydrogen, halogens, hydrogen halides and steam with alkenes
- describe the mechanism of electrophilic addition to alkenes
- describe addition polymerisation, by drawing the polymer obtained from a given monomer
- discuss the problems of disposal of polymers

# Additional reading

The search for fuels to replace oil-based products is gathering momentum not only because the Earth's reserves will run out (perhaps during this century, if current consumption is maintained) but because of the need to limit air pollution. Many countries wish to reduce their dependence on oil-producing nations and, ideally, also reduce costs. Alternatives that fulfil these objectives are not easy to find, particularly because huge quantities of fuel are required.

Fuels often require specific chemical properties. The fuel in cars has to be volatile enough to allow the engine to start in cold conditions, but not so volatile as to be dangerous in hot conditions. It should burn cleanly and vigorously during acceleration and run smoothly when cruising. Indeed, the isomerisation and reforming processes in the oil industry provide mixtures of components that optimise these properties. Alternatives that meet some of the required objectives include ethanol, methanol, biodiesel, liquefied petroleum gas (LPG) and hydrogen. There are no easy answers and research is dedicated to establishing a way forward.

In Brazil, ethanol derived from sugar cane is widely used as a fuel. Ethanol can also be made from corn. It is a promising alternative, in that it is carbon neutral. However, even if all corn production were used for this purpose, the amount of ethanol produced would fall short of current demand. Other crops and organic waste might be a partial answer and a number of different lines of research are being followed. In particular, cellulose-based materials such as grasses have been utilised, which has succeeded in reducing production costs.

Another potential fuel is methanol, which can be obtained from coal in a two-stage process that requires suitable catalysts:

$$C(s) + H_2O(g) \rightarrow CO(g) + H_2(g)$$
$$CO(g) + 2H_2(g) \rightarrow CH_3OH(g)$$

This is not ideal as coal is a non-renewable fossil fuel and there are pollution problems associated with its use. It contains sulphur, which is oxidised to sulphur dioxide, although this could be removed by absorption into an alkali. Coal has an empirical formula of around $C_{240}H_{90}O_4S$, though this is variable.

It may be better to use carbon dioxide as the starting material for methanol production:

$$CO_2(g) + 3H_2(g) \rightarrow CH_3OH(g)$$

This has the advantage of using up unwanted carbon dioxide produced by industrial processes — for example, the manufacture of lime. The hydrogen gas could be generated by the electrolysis of water, but this process requires energy that, on a large scale, could be provided only by nuclear power. However, methanol is in many ways an ideal fuel. Current car engines would not need significant adaptation. Like ethanol, it has the disadvantage of low volatility, which means that cold-starting the engine is problematic. However, used in combination with 15% petrol, methanol has been shown to be an effective fuel.

Biodiesel made from soybean oil, waste cooking oil or rapeseed oil is a potentially cheap fuel. When burned, it produces less carbon monoxide and unburned particles than petrol but releases more nitrogen oxides. This is a concern, because not only are nitrogen oxides poisonous, they are also components of chemical smog and, in the upper atmosphere, attack the ozone layer.

In terms of reducing pollution, hydrogen is the ideal fuel because its combustion product is water. At present, much hydrogen is made from natural gas, but it can also be obtained from the electrolysis of water, which requires a large energy input. It is important to consider the downside of this, before hailing hydrogen as the best way forward. There is a further problem in that using hydrogen as a fuel increases the likelihood of more nitrogen oxide emissions because hydrogen burns at a higher temperature.

LPG is efficient, but its source is mainly oil deposits and underground reserves. Therefore, it may not be a long-term solution to the supply of fuel. LPG does have the advantage of producing smaller quantities of polluting gases than petrol does.

There is not a single solution to the future need for fuels; a mixture of technologies is likely to be used. Hybrid vehicles (partly powered by electricity) are already available and the use of fuel cells will develop rapidly in the near future.

## Extension question: what is the fuel?

The action of the bacterium *Clostridium aceto-butylicum* on biomass, such as sugar beet, wheat or straw, is used to produce compound X, which is a fuel. The fuel, X, can produce almost as much power as petrol, although it cannot be used in a pure form because it exists as a gel below 25.5°C.

A sample of X contained the following elements by mass: carbon 64.9%, hydrogen 13.5%, oxygen 21.6%.

a  Calculate the empirical formula of X.

When X is vaporised, it is found that 0.2g of X has a volume of 65 cm³ measured at room temperature and pressure.

b  Calculate the molecular formula of X.

Compound X reacts with concentrated sulphuric acid and a hydrocarbon, Y, is formed that contains the same number of carbon atoms as X.

When bubbled into aqueous bromine, the bromine is decolorised.

c  Compound Y could be one of four isomers. Draw the displayed formulae of these isomers.

d  Choose one of the isomers and write a balanced equation for its reaction with bromine.

Compound Y reacts with hydrogen to give an unbranched hydrocarbon, Z.

e  Give the conditions for the reaction of compound Y with hydrogen.

f  Name hydrocarbon Z.

When compound Y reacts with steam, two products are obtained, one of which is X.

g  Give the structural formulae of the two products.

h  The functional group of X is at the end of its carbon chain. Give the name of compound X.

# Alcohols, halogenoalkanes and analysis

## Chapter 10

This chapter covers the reactions of two more homologous series: alcohols and halogenoalkanes. Alcohols undergo a wider variety of reactions than hydrocarbons. Halogenoalkanes react by nucleophilic substitution.

Most analysis of organic molecules is now carried out using instrumental methods. For example, infrared spectroscopy is used to identify the presence of particular bonds within a molecule, by the energy with which they vibrate. Mass spectrometry provides a detailed analysis of the structure of atoms and molecules by recording their masses. The use of both of these techniques is described.

## Alcohols

A molecule of an alcohol contains the hydroxy group, –OH. The names of all alcohols end with -ol. Molecules of many naturally occurring compounds, such as carbohydrates, pheromones, vitamins and steroids, contain an –OH group.

Menthol is a naturally occurring alcohol that is extracted from mint oils. It has local anaesthetic and anti-irritant properties and is used to relieve minor throat irritation.

TOPFOTO

*Peppermint is a common garden herb and a source of mint oils*

# Classification of alcohols

All alcohols contain a carbon atom bonded to a hydroxy group, C–OH. They are classified as primary, secondary or tertiary alcohols:

- In a **primary alcohol**, the carbon atom bonded to the –OH group is bonded to only one other carbon atom:

- In a **secondary alcohol**, the carbon atom bonded to the –OH group is bonded to two carbon atoms:

- In a **tertiary alcohol**, the carbon atom bonded to the –OH group is bonded to three carbon atoms:

Examples of each type of alcohol are shown in Figure 10.1.

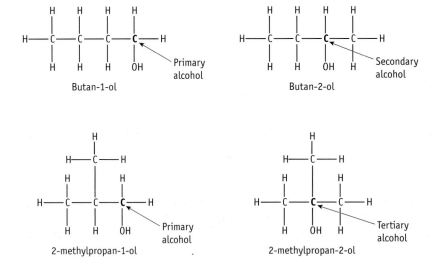

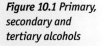

*R, R′ and R″ are used to represent groups other than –H.*

*Figure 10.1 Primary, secondary and tertiary alcohols*

# Physical properties

Alcohols have relatively high boiling and melting points and are miscible with water. These factors can be explained in terms of hydrogen bonding (page 92). A hydrogen bond forms between the lone pair of electrons on the oxygen in the –OH group in one alcohol molecule and the hydrogen in the –OH group in an adjacent alcohol molecule:

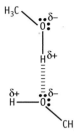

### Boiling point and melting point

When alcohols are boiled (or melted), energy is required to break the inter-molecular forces. All substances, including alcohols, have van der Waals forces that help bind the molecules together. In addition, alcohol molecules have hydrogen bonds. Hydrogen bonds are the strongest of the intermolecular forces, so more energy is required to break them than to overcome van der Waals forces. Hydrogen bonding decreases the volatility and, therefore, results in an increase in boiling point.

### Miscibility with water

Methanol and ethanol are freely miscible with water. When mixed, some of the hydrogen bonds in the individual liquids are broken, but they are then replaced by new hydrogen bonds between the alcohol and water:

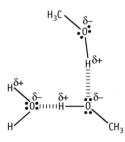

As the relative molecular mass of the alcohol increases, the miscibility with water decreases.

# Preparation of ethanol

Ethanol is widely used as a solvent. Most industrial ethanol is made by the addition reaction of steam with ethene in the presence of a phosphoric acid, $H_3PO_4$, catalyst (page 160). This a hydration reaction:

$$CH_2{=}CH_2(g) + H_2O(g) \rightarrow CH_3{-}CH_2OH(g)$$

**e** Remember that a hydration reaction is not the same as hydrolysis.

Ethene is a co-product in the cracking of long-chain alkanes (pages 146–47).

Ethanol can also be produced by anaerobic fermentation of sugar found in fruit and grain. Fermentation involves yeasts, which occur naturally. The process is carried out in the absence of oxygen (anaerobic conditions) and glucose (a sugar) is converted to ethanol and carbon dioxide:

$$C_6H_{12}O_6(aq) \rightarrow 2CH_3CH_2OH(aq) + 2CO_2(g)$$

## Uses of alcohols

- Ethanol has been known to humans for thousands of years and is the active component in alcoholic drinks.
- Industrial alcohol, in the form of methylated spirits, is used as a solvent.
- Ethanol is being used increasingly as a biofuel and as a petrol substitute in countries with limited oil reserves.
- Methanol is also used as a petrol additive to improve combustion and is an important feedstock in the production of organic chemicals.

## Chemistry of alcohols

The products of the reactions of alcohols, such as aldehydes, ketones, carboxylic acids and esters, contain new functional groups. It is important that you are able to recognise and name these compounds and their functional groups. However, they are covered in greater depth at A2.

### Combustion

Alcohols burn to produce carbon dioxide and water. The reactions are exothermic, releasing large amounts of energy:

$$CH_3OH + 1\tfrac{1}{2}O_2 \rightarrow CO_2 + 2H_2O$$
$$C_2H_5OH + 3O_2 \rightarrow 2CO_2 + 3H_2O$$

### Elimination

When reacted with hot, concentrated sulphuric acid or hot pumice/$Al_2O_3$, an alcohol is dehydrated to form an alkene.

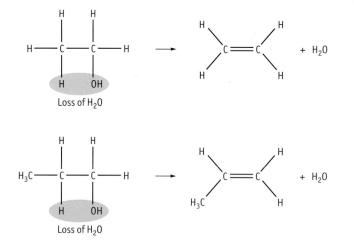

When balancing the equation, don't forget to include the oxygen in the alcohol.

For alcohols such as butan-2-ol, water can be lost in two ways:

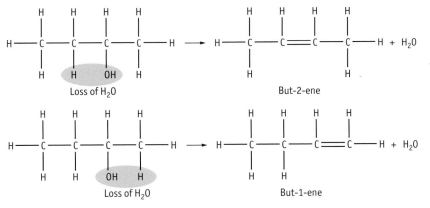

The C=C bond is formed by an **elimination reaction**. In this case, the water is eliminated and the reaction could also be described as a **dehydration reaction**.

### Esterification

When an alcohol reacts with a carboxylic acid in the presence of an acid catalyst, an ester is formed.

A **carboxylic acid** contains the functional group:

An **ester molecule** contains a link that joins groups from the acid and the alcohol together:

Esters are named from both the alcohol and the carboxylic acid. The first part of the name relates to the alcohol and the second part comes from the carboxylic acid. If methanol reacts with ethanoic acid, the ester is methyl ethanoate (methyl from methanol; ethanoate from ethanoic acid):

$$CH_3COOH \ + \ CH_3OH \ \rightleftharpoons \ CH_3COOCH_3 \ + H_2O$$
**Ethanoic acid    Methanol     Methyl ethanoate**

In an organic reaction, it is the functional groups that react. In this particular reaction it is helpful to write the formulae so that the functional groups face each other. The alcohol reacts with carboxylic acid to produce water and the ester, which can be deduced by simply joining the two organic parts together:

$$CH_3COOH + HOCH_3 \rightleftharpoons CH_3COOCH_3 + H_2O$$

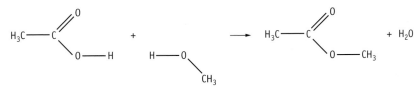

> **e** Both the
> *E*- and *Z*-versions
> of but-2-ene can
> be formed.

There are two ways in which the ester may be formed because there are two different bonds in the alcohol that could break. The bond between the hydrogen and the oxygen might break:

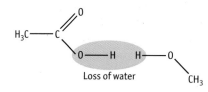

The bond between the carbon and the oxygen could also break:

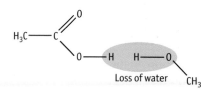

The mechanism has been studied using alcohols containing the $^{18}O$ isotope. The two possible mechanisms are:

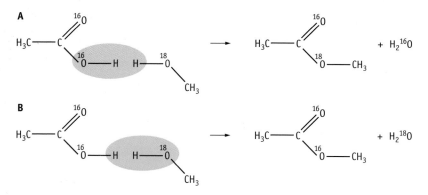

Analysis of the product ester by mass spectrometry (page 189) shows that the $^{18}O$ is found in the ester and not in the water. Therefore, the reaction proceeds as shown in mechanism A above.

A range of esters can be formed by changing either the alcohol or the carboxylic acid, or both — for example:

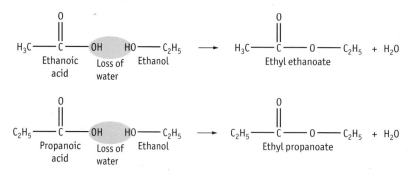

### Properties of esters

Esters have a characteristic smell that is described as 'fruity'. They are used in perfumes and artificial flavourings. Naturally occurring fruits contain a complex mix of chemicals, many of which contribute to the overall scent. By mixing together different chemicals, many of which are esters, it is possible to manufacture artificial flavours. The esters shown in Figure 10.2 are used to generate the flavours pineapple, pear and apple.

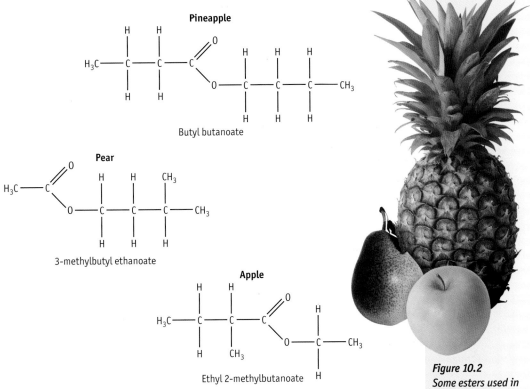

**Pineapple**

Butyl butanoate

**Pear**

3-methylbutyl ethanoate

**Apple**

Ethyl 2-methylbutanoate

*Figure 10.2*
*Some esters used in artificial flavourings*

### Oxidation

When alcohols are warmed with an oxidising mixture such as acidified dichromate, $Cr_2O_7^{2-}/H^+$ (e.g. $K_2Cr_2O_7/H_2SO_4$), oxygen atoms are added to their structure. For this reaction, the extent to which this occurs depends on the type of alcohol:

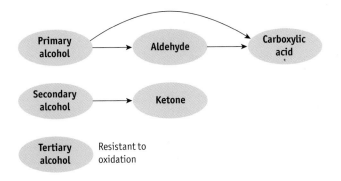

Primary alcohol → Aldehyde → Carboxylic acid

Secondary alcohol → Ketone

Tertiary alcohol — Resistant to oxidation

An **aldehyde** contains the functional group:

A **ketone** looks similar, but the functional group is:

Balanced equations for the oxidation of alcohols can be written using [O] to represent the oxidising agent. The equations for the oxidation of a **primary alcohol** to an **aldehyde** or a **secondary alcohol** to a **ketone** are similar in that water is always formed:

- primary alcohol + [O] → aldehyde + water
- secondary alcohol + [O] → ketone + water

*Oxidation of a primary alcohol to an aldehyde*
Methanol is oxidised to methanal:

$$CH_3OH + [O] \rightarrow HCHO + H_2O$$
Methanol   Methanal

Ethanol is oxidised to ethanal:

$$CH_3CH_2OH + [O] \rightarrow CH_3CHO + H_2O$$
Ethanol    Ethanal

*Oxidation of a secondary alcohol to a ketone*
Propan-2-ol is oxidised to propanone:

$$CH_3CH(OH)CH_3 + [O] \rightarrow CH_3COCH_3 + H_2O$$
Propan-2-ol    Propanone

Pentan-2-ol is oxidised to pentan-2-one:

$$CH_3CH_2CH_2CH(OH)CH_3 + [O] \rightarrow CH_3CH_2CH_2COCH_3 + H_2O$$
Pentan-2-ol     Pentan-2-one

*Further oxidation of a primary alcohol to a carboxylic acid*
After an aldehyde has been formed by the oxidation of a primary alcohol, a further oxidation step can take place. The product is a **carboxylic acid**. Balanced equations for the oxidation of a primary alcohol to a carboxylic acid can be represented as:

primary alcohol + 2[O] → carboxylic acid + water

Methanol can be oxidised to methanoic acid:

$$CH_3OH + 2[O] \rightarrow HCOOH + H_2O$$
Methanol   Methanoic acid

Ethanol can be oxidised to ethanoic acid:

$$CH_3CH_2OH + 2[O] \rightarrow CH_3COOH + H_2O$$
Ethanol    Ethanoic acid

### Reasons for the different behaviour of the alcohols in oxidation

You should notice an important point about the oxidation of alcohols. Oxidation occurs readily only if the carbon attached to the hydroxy group is also bonded to an H. Both primary and secondary alcohols contain the necessary –CHOH grouping, but tertiary alcohols do not. Oxidation of a tertiary alcohol can be achieved only if a C–C bond is broken. It is generally very difficult to break a C–C bond in chemical reactions. It happens during combustion but rarely elsewhere.

The C–O–H link in a tertiary alcohol molecule cannot be converted to a C=O group (the first step in the oxidation) without one of the three C–C bonds attached to the –OH group in the molecule being broken. Therefore, tertiary alcohols resist oxidation.

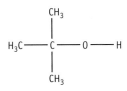

A secondary alcohol molecule contains a C–H bond. A secondary alcohol can, therefore, be oxidised because a C=O unit can be created without a C–C bond having to be broken.

However, the ketone that is produced cannot be oxidised further.

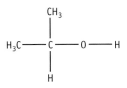

A primary alcohol molecule contains two C–H bonds. A primary alcohol can, therefore, be oxidised because a C=O unit can be created without breaking a C–C bond.

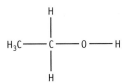

### Apparatus used in the oxidation of alcohols

A primary alcohol can be oxidised in two steps. The first step creates the aldehyde. In the next step, the second C–H is converted into C–O–H so that the carboxylic acid functional group is formed.

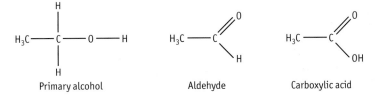

Primary alcohol          Aldehyde          Carboxylic acid

When oxidising a primary alcohol, the choice of apparatus is important. The alcohol and the oxidising mixture have to be heated. This can be achieved by using either apparatus for distillation (Figure 10.3) or a reflux condenser (Figure 10.4).

If ethanol is oxidised, either ethanal or ethanoic acid can be produced. The boiling point of:

- ethanol, $CH_3CH_2OH$, is 78°C
- ethanal, $CH_3CHO$, is 21°C
- ethanoic acid, $CH_3COOH$, is 118°C

The product depends on the apparatus used.

*Oxidation of ethanol to ethanal*

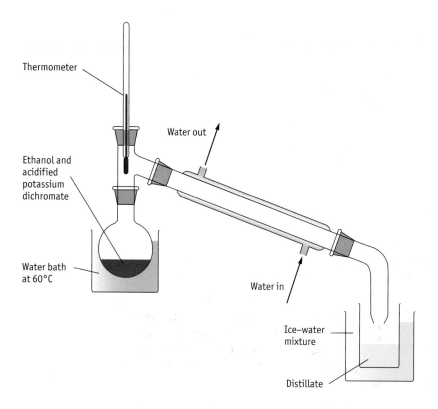

Thermometer

Water out

Ethanol and acidified potassium dichromate

Water bath at 60°C

Water in

Ice–water mixture

Distillate

*Figure 10.3
Distillation
apparatus for
the oxidation of
a primary alcohol
to an aldehyde*

The distillation process involves evaporation followed by condensation, which allows the most volatile component to be separated. When ethanol is heated with a mixture of potassium dichromate(VI) and sulphuric acid in a distillation apparatus (Figure 10.3), the volatile component evaporates first. Ethanal has the lowest boiling point and, therefore, vaporises most readily. The condenser has cold water circulating in the outer sleeve, so when the ethanal reaches the condenser, it condenses, is separated from the reaction mixture, and can be collected.

During the oxidation reaction, there is a distinctive colour change from orange to green. This occurs because the orange dichromate(VI) ion, $Cr_2O_7^{2-}$, present in the oxidising mixture is reduced to green chromium(III).

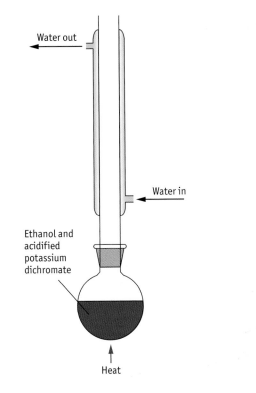

Water out

Water in

Ethanol and acidified potassium dichromate

Heat

*Figure 10.4 Reflux apparatus for the oxidation of a primary alcohol to a carboxylic acid*

*Oxidation of ethanol to ethanoic acid*
Reflux involves a process of continuous evaporation and condensation, which prevents volatile components from escaping.

As with the distillation, when the reaction mixture is heated the most volatile component vaporises first. Ethanal has the lowest boiling point and is vaporised most readily. The vertical reflux condenser has cold water circulating in the outer sleeve. When the ethanal reaches the condenser, it condenses and falls back into the oxidising mixture. Therefore, it is not separated from the reaction mixture. The ethanal is oxidised further to form ethanoic acid.

As with the oxidation of ethanol to ethanal, the production of ethanoic acid is accompanied by a colour change from orange to green as the orange dichromate(VI) ion, $Cr_2O_7^{2-}$, is reduced to green chromium(III).

## Diols and triols

A molecule of any of the alcohols discussed so far contains just one hydroxy group (–OH). A **diol** contains two –OH groups and a **triol** contains three –OH groups. Diols and triols behave similarly to monohydric alcohols, but the properties are accentuated.

**Monohydric alcohols**

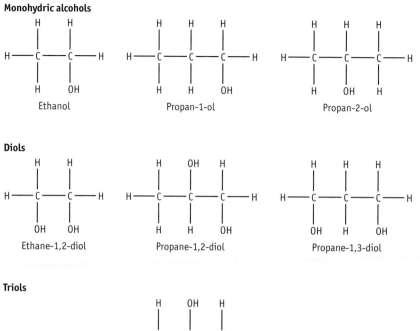

Ethanol  Propan-1-ol  Propan-2-ol

**Diols**

Ethane-1,2-diol  Propane-1,2-diol  Propane-1,3-diol

**Triols**

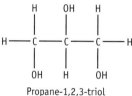

Propane-1,2,3-triol

Diols and triols form more hydrogen bonds than monohydric alcohols because they contain more –OH groups. This increase in hydrogen bonding leads to increases in viscosity, density and boiling point. The densities and boiling points of some alcohols are shown in Table 10.1.

| | Density/g cm$^{-3}$ | Boiling point/°C |
|---|---|---|
| Ethanol | 0.79 | 78 |
| Ethane-1,2-diol | 1.11 | 198 |
| Propan-1-ol | 0.81 | 97 |
| Propane-1,3-diol | 1.19 | 215 |
| Propane-1,2,3-triol | 1.26 | 290 |

The chemical properties of diols and triols are similar to those of monohydric alcohols. However, they may react only partially. More vigorous conditions are sometimes required for the reaction of the second or third –OH group. For example, ethane-1,2-diol reacts with ethanoic acid to form an ester:

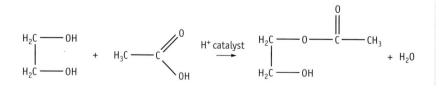

*Figure 10.5*
*Alcohols containing one, two or three hydroxy groups*

*Table 10.1*
*Densities and boiling points of alcohols*

However, if the carboxylic acid is in excess, a diester is formed:

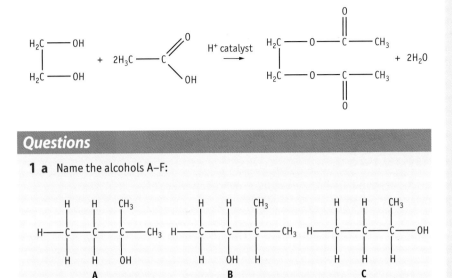

## Questions

**1 a** Name the alcohols A–F:

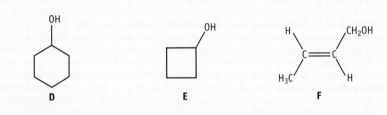

**b** Answer the following questions about compounds A–F in part **a**:
   **(i)** What is the molecular formula of compound D?
   **(ii)** Identify any secondary alcohols.
   **(iii)** What do compounds E and F have in common?

**2** Write a balanced equation for the following reactions:
  **a** the complete combustion of propan-1-ol
  **b** the dehydration of pentan-3-ol
  **c** the reaction between butane-1,3-diol and excess methanoic acid
  **d** the oxidation of butan-2-ol (use [O] to represent the oxidising agent)

**3** Explain, with the aid of equations, why the dehydration of pentan-3-ol gives two alkenes, whereas the dehydration of pentan-2-ol gives a mixture of three alkenes.

**4** Explain what is meant by the terms 'reflux' and 'distillation'.

**5** Identify the organic products formed when 2-methylcyclohexanol undergoes dehydration.

**6 a** Explain why ethane-1,2-diol has a much higher boiling point than ethanol.
   **b** Write a balanced equation for the complete combustion of ethane-1,2-diol.
   **c** Using [O] to represent the oxidising agent, write a balanced equation for the complete oxidation of ethane-1,2-diol.
   **d** Propane-1,2,3-triol can be oxidised to form a mixture of organic products. When oxidised completely, HOOCCOCOOH is formed. Draw the displayed formula of HOOCCOCOOH and identify as many of the partial oxidation products as possible.

# Halogenoalkanes

A halogenoalkane is a compound in which one or more hydrogen atoms of an alkane is replaced by a halogen atom.

If one hydrogen is replaced, the general formula is $C_nH_{2n+1}X$ (where X = F, Cl, Br or I).

Like alcohols, halogenoalkanes are subdivided into primary, secondary and tertiary. The rules for classification are the same. If the carbon atom that is bonded to the halogen (X) is bonded to one carbon atom only, then the compound is a primary halogenoalkane. If the carbon atom in the C–X bond is bonded to two other carbon atoms, the compound is a secondary halogenoalkane; if it is bonded to three other carbon atoms, the compound is a tertiary halogenoalkane.

An alternative way of recognising whether halogenoalkanes are primary, secondary or tertiary is as follows:

- A primary halogenoalkane contains the group:

- A secondary halogenoalkane contains the group:

- A tertiary halogenoalkane contains the group:

The isomers of $C_4H_9Cl$ are classified thus:

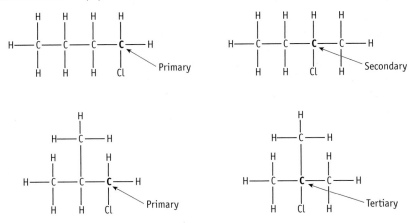

Primary, secondary and tertiary halogenoalkanes behave similarly. They react with the same reagents, but the rate at which they react differs. The rate of reaction depends on the strength of the C–X bond. Tertiary halogenoalkanes react faster than secondary halogenoalkanes; primary halogenoalkanes react most slowly. This indicates that the C–X bond in a tertiary halogenoalkane is the weakest and the C–X bond in a primary halogenoalkane is the strongest.

## Nucleophilic substitution in halogenoalkanes

A nucleophile is an electron-pair donor. The electrons donated are a lone pair on an electronegative atom.

The carbon–halogen bond is polar. The charge separation leaves the carbon atom open to attack by a nucleophile.

Nucleophilic substitution of a primary halogenoalkane is described by the following general mechanism:

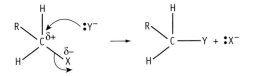

The initial attack by the nucleophile involves the lone pair of electrons on the $Y^-$ ion forming a new bond with the carbon and, simultaneously, the bonded pair of electrons in the C–X bond breaking away to form a lone pair of electrons on the resulting $X^-$ ion.

Halogenoalkanes are attacked by a range of nucleophiles. However, it is sufficient to understand the mechanism and conditions for a hydrolysis reaction only.

### Hydrolysis
When a primary halogenoalkane is heated under reflux with an aqueous solution of an alkali (e.g. sodium hydroxide or potassium hydroxide), the halogenoalkane is hydrolysed to a primary alcohol.

**e** A curly arrow represents the movement of an electron pair.

**Reagent:** sodium (or potassium) hydroxide (an alkali is required)

**Conditions:** the solvent *must* be water and the reaction mixture *must* be heated under reflux

**Equation:** $CH_3CH_2Br + NaOH \rightarrow CH_3CH_2OH + NaBr$

**Mechanism:**

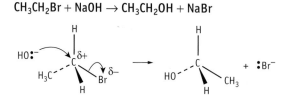

The hydrolysis reaction can be monitored in one of two ways.

### Method 1

The halogenoalkane is reacted with excess sodium hydroxide for a fixed period of time. The excess sodium hydroxide is then neutralised by the addition of nitric acid, $HNO_3$. Aqueous silver nitrate, $AgNO_3$, is added and the density of the precipitate is monitored. The experiment is repeated for other fixed periods of time.

### Method 2

The halogenoalkane is mixed with water and a small amount of aqueous silver nitrate solution containing some ethanol in a water bath at 60°C. The water can act as a nucleophile (using a lone pair of electrons on its oxygen atom), but the reaction is slow. As the hydrolysis reaction occurs, the halide ion is displaced and reacts with the $Ag^+$ ions to give a precipitate. By monitoring the rate at which the precipitate appears, it is possible to deduce the rate of hydrolysis. The ethanol is added to aid better mixing of the reagents since, although halogenoalkanes and water are immiscible (i.e. they form separate layers when mixed), they both dissolve in ethanol. The ethanol may also be involved as a nucleophile. If a comparison has to be made between halogenoalkanes, the conditions have to be carefully controlled.

### Reactivity of halogenoalkanes

When 1-chlorobutane, 1-bromobutane and 1-iodobutane are reacted under identical conditions, 1-iodobutane reacts the fastest and 1-chlorobutane reacts the slowest. This may seem surprising, since the charge separation is greatest for C–Cl and least for C–I — the nucleophile has the strongest attraction to the carbon atom in a chloroalkane. However, the other factor that determines the rate of the reaction is the strength of the carbon–halogen bond. Bond strengths (enthalpies) are listed in Table 10.2.

| Bond | Bond enthalpy/kJ mol$^{-1}$ |
|------|------------------------------|
| C–F | 467 |
| C–Cl | 340 |
| C–Br | 280 |
| C–I | 240 |

*Table 10.2*
*Bond enthalpies of carbon–halogen bonds*

The C–I bond is the weakest and the least energy is required to break it. The C–F bond is so strong that it rarely undergoes hydrolysis.

The strength of the carbon–halogen bond is the dominant factor in determining rates of reaction of halogenoalkanes. For example, when 1-chlorobutane, 2-chlorobutane and 2-chloro-2-methylpropane are reacted under identical conditions, 2-chloro-2-methylpropane reacts most quickly and 1-chlorobutane reacts most slowly. This can be explained by comparing the relevant carbon–chlorine bond enthalpies (Table 10.3).

The C–Cl bond in the tertiary chloroalkane is the weakest and, therefore, requires the least amount of energy to break it. The C–Cl bond in the primary chloroalkane is the strongest, so the primary chloroalkane is the most difficult to hydrolyse.

| Bond | Bond enthalpy/kJ mol$^{-1}$ |
|------|------|
| C–Cl (primary) | 340 |
| C–Cl (secondary) | 331 |
| C–Cl (tertiary) | 289 |

*Table 10.3*
*Bond enthalpies in primary, secondary and tertiary chloroalkanes*

## Uses of halogenoalkanes

Halogenoalkanes are used in the preparation of a wide range of products, including pharmaceuticals (such as ibuprofen) and polymers (such as PVC and PTFE). Halogenoalkanes are also used in industry as solvents and lubricants.

Table 10.4 shows four complex halogenated compounds and their uses, other than in the manufacture of the plastics PVC and PTFE (page 163).

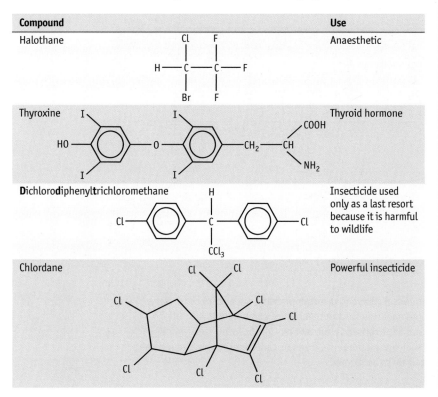

| Compound | | Use |
|----------|--|-----|
| Halothane | | Anaesthetic |
| Thyroxine | | Thyroid hormone |
| Dichlorodiphenyltrichloromethane | | Insecticide used only as a last resort because it is harmful to wildlife |
| Chlordane | | Powerful insecticide |

*Table 10.4*
*Complex halogenated compounds and their uses*

## CFCs

Halogenoalkanes were used to produce chlorofluorocarbons (CFCs), such as dichlorodifluoromethane, $CCl_2F_2$, and trichlorofluoromethane, $CCl_3F$. Their role in the destruction of the ozone layer is considered in detail on pages 231–32.

The use of CFCs is being phased out. However, it is worth appreciating why they were originally chosen for use in air conditioning, refrigeration units and aerosols. These applications require a liquid of suitable volatility that can be readily evaporated and re-condensed. In addition, the liquid must be unreactive, non-flammable and non-toxic. CFCs fit these criteria and they were, therefore, considered to be safe and environmentally friendly. However, it is these properties that make them so persistent in the atmosphere. At the time of their introduction, the dangerous effect they would have in the stratosphere was not understood. Once this was appreciated, alternatives were sought.

Initially, alternatives centred on HCFCs, such as 1,1,1,2-tetrafluoro-2-chloro-ethane, which have a C–H bond that makes them more degradable in the atmosphere. However, these have also been found to be unsatisfactory because some HCFCs do persist and reach the stratosphere, releasing chlorine as a radical. Current thinking is to use hydrocarbons, but much effort is being put into developing HFCs. Like HCFCs, HFCs contain a C–H bond and therefore degrade in the atmosphere. However, HFCs contain no chlorine so they do not directly affect stratospheric ozone, but the products of HFC degradation may have other harmful environmental effects.

## Questions

**1 a** Name the halogenoalkanes A–F:

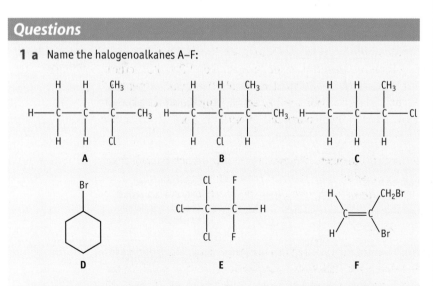

**b** Classify compounds A, B and C as either primary, secondary or tertiary.

**c** Write a balanced equation for the reaction of compound D with $OH^-$.

**d** When compound E is exposed to ultraviolet light it forms free radicals. Explain what is meant by the term 'free radical'. Identify the free radicals that are most likely to be formed.

**2** The structure of chlordane is shown below. What are the molecular and empirical formulae of chlordane?

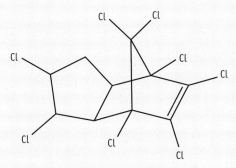

**3** Halogenoalkanes undergo nucleophilic substitution reactions.
  **a** Define a nucleophile.
  **b** Explain fully the mechanism when 1-bromopropane reacts with potassium hydroxide, KOH, to produce propan-1-ol.
  **c** Explain fully the mechanism when 2-bromo-2-methylbutane reacts with potassium hydroxide, KOH, to produce 2-methylbutan-2-ol.

# Modern analytical techniques

## Infrared spectroscopy

Substances used to be identified by means of distinctive chemical tests. In simple cases, such tests are still useful. However, more powerful and flexible methods have been developed that use instruments to detect molecules by their structural characteristics. One such method is **infrared spectroscopy**. Infrared radiation causes covalent bonds to vibrate; particular bonds respond at different frequencies (energies).

Infrared radiation is part of the electromagnetic spectrum, which extends from low-energy waves, such as radio waves, through to high-energy waves, such as X-rays and gamma rays. Infrared is the region of the spectrum that has energy just below that of red light.

### Absorptions

Covalent bonds respond to infrared radiation in a number of ways. The frequencies at which the bonds vibrate are called the **absorptions** of that bond. Some absorptions occur over a small range of frequencies (such as those of the C=O bond); others have a wider spread (such as those of an O–H bond). The absorptions are affected by neighbouring bonds. For example, the O–H bond of an alcohol responds differently from the O–H bond in a carboxylic acid group, –COOH. The complicated patterns that are produced can make the interpretation of these spectra difficult.

There is no need to consider here how an infrared spectrometer works, but you should be able to recognise some key absorptions (Table 10.5). Absorptions are identified by **wavenumber** (cm$^{-1}$).

| Wavenumber range/cm$^{-1}$ | Bond | Functional groups |
|---|---|---|
| 2500–3300 (very broad) | O–H | Carboxylic acid |
| 3200–3550 | O–H | Alcohol |
| 1640–1750 | C=O | Aldehyde, ketone, carboxylic acid, ester |
| 1000–1300 | C–O | Alcohol, ester, carboxylic acid |

*Table 10.5*
*Key infrared absorptions*

**e** Data sheets are supplied when required in exams, so there is no need to learn the wavenumbers in Table 10.5.

### Infrared spectrum of an alcohol

The identifying absorptions are indicated on the spectrum. Notice the characteristic broad absorption due to the O–H bond in the region 3200–3550 cm$^{-1}$.

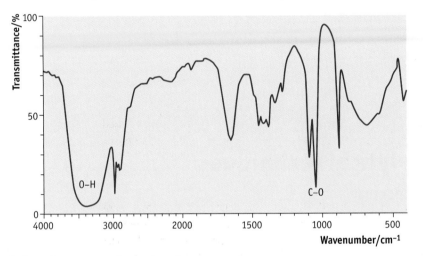

### Infrared spectrum of a carbonyl compound

The infrared spectrum of a carbonyl compound has an absorption due to the C=O bond in the region 1640–1750 cm$^{-1}$.

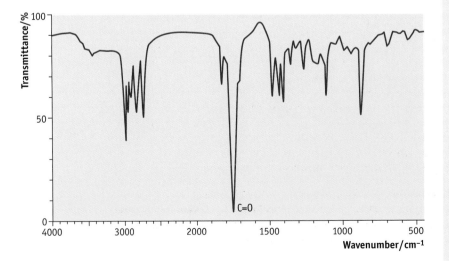

*Infrared spectrum of a carboxylic acid*

The infrared spectrum of a carboxylic acid has absorptions due to:

- the C=O bond (in the region 1640–1750 cm$^{-1}$)
- the O–H bond, which is a very broad absorption (in the region 2500–3500 cm$^{-1}$)

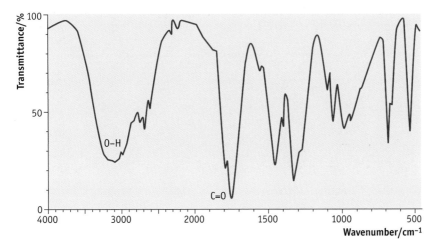

The above spectra should make it easy to identify compounds that contain the O–H and C=O functional groups and they also illustrate the use of infrared spectroscopy. Modern breathalysers employ this technique quantitatively to measure the amount of ethanol in exhaled breath.

## Mass spectrometry

The use of the mass spectrometer to determine the relative isotopic masses of elements was discussed in Chapter 1. It is important to remember that the mass spectrometer analyses *positive* ions only. The positive ions are generated by the interaction of high-energy electrons with a gaseous chemical sample. The high-energy electrons remove an electron from the sample. This is represented by the equation below, where the sample chemical is X:

$$e^- + X(g) \rightarrow X^+(g) + 2e^-$$

### Analysis of organic compounds

The mass spectrometer can also analyse compounds, although the number of lines obtained is usually large. This is because the bombardment by electrons causes the molecule to break up and each of the positively charged fragments obtained registers on the detector. This can be an advantage, as it is sometimes possible to obtain details of the structure of the molecule as well as its overall molecular mass.

*Mass spectrum of propane*

The mass spectrum of propane, $C_3H_8$, is shown in Figure 10.6.

The relative molecular mass of propane is 44. The peak furthest to the right of the spectrum represents the ion $C_3H_8^+$. This is called the **molecular-ion peak**. The molecular ion is produced by the removal of an electron from the molecule:

$$e^- + H_3C–CH_2–CH_3(g) \rightarrow (H_3C–CH_2–CH_3)^+(g) + 2e^-$$

However, the molecular ion is unstable and breaks down to ion fragments of the molecule, which are also detected. Some examples are labelled on Figure 10.6. The peak at $m/z$ 29 occurs because a C–C bond in the $CH_3CH_2CH_3$ chain of the molecular ion is broken, producing two fragments. One of the fragments has a positive charge and the other is neutral:

$$(H_3C-CH_2-CH_3)^+(g) \rightarrow (CH_3-CH_2)^+(g) + CH_3(g)$$

In this case, the ion $CH_3CH_2^+$ is detected.

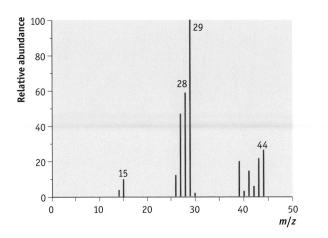

**Figure 10.6**
*Mass spectrum of propane, $C_3H_8$*

ⓔ Remember that it is positive ions that are detected by a mass spectrometer.

The peak at $m/z$ 15 represents a $CH_3^+$ ion. Again, a C–C bond in the $CH_3CH_2CH_3$ chain is broken, producing one fragment with a positive charge and one fragment that is neutral:

$$(H_3C-CH_2-CH_3)^+(g) \rightarrow CH_3^+(g) + CH_3-CH_2(g)$$

A fragment can be identified for all other peaks in the spectrum. It is because of this fragmentation that the set of peaks is often described as a **fingerprint** of the molecule.

### Mass spectrum of ethanol
The mass spectrum of ethanol is shown in Figure 10.7.

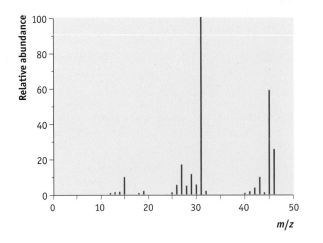

**Figure 10.7** *Mass spectrum of ethanol*

The molecular ion is represented by the line at $m/z$ 46. This is the line furthest to the right, but it is not the highest peak. The height of the line represents the relative amount of that ion formed from the sample. It is often the case that a molecule breaks down readily, rather than surviving as the molecular ion. The peak at $m/z$ 45 is due to the loss of H from the –OH group, resulting in the ion $CH_3CH_2O^+$ (g). The most abundant peak is at $m/z$ 31, which corresponds to the ion $CH_2OH^+$ (g), formed by the fragmentation of ethanol by breaking the C–C bond.

### Structural isomers

Analysis of the fragmentation of molecules can be used to distinguish between structural isomers. Computers are used to compare the pattern obtained with a database of spectra. This is a widely used research tool.

### Lines due to isomers

The precision of a mass spectrometer is such that lines due to isotopes can be observed (not shown in Figure 10.6 and Figure 10.7). These lines have small peak heights. One characteristic is the presence of such small peaks beyond the molecular ion peak. These are usually the result of small numbers of ions containing the isotope $^{13}C$, although other isotopes may contribute. These peaks are excluded when using a mass spectrum to identify a compound.

# Summary

You should now be able to:
- explain the low volatility of alcohols and their solubility in water
- describe the preparation of ethanol by fermentation of a sugar with yeast (enzymes) and by the reaction of steam with ethene
- explain the classification of alcohols as primary, secondary or tertiary
- describe, and write equations for, the combustion, dehydration and esterification of alcohols
- describe, and write equations for, the oxidation of primary and secondary alcohols and relate these to their structures
- explain the resistance of tertiary alcohols to oxidation
- describe the mechanism of nucleophilic substitution for halogenoalkanes and illustrate this using hydrolysis
- explain the ease of hydrolysis of the iodo-, bromo- and chloroalkanes
- outline some uses of halogenoalkanes and understand why CFCs were chosen originally
- identify –OH, C=O and –COOH groups using infrared spectroscopy
- understand how mass spectrometry can be used to determine relative isotopic mass
- interpret a mass spectrum to provide a relative mean atomic mass (Chapter 1)
- use the molecular-ion peaks to interpret lines in the mass spectra of molecules

# Additional reading

When ethane-1,2-diol is esterified with a dibasic carboxylic acid, a polymer is formed (Figure 10.8). The polymer is a polyester and has a wide range of everyday uses.

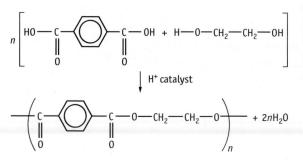

$\downarrow$ H$^+$ catalyst

$+ 2nH_2O$

*Figure 10.8*

> **e** This type of polymerisation is studied in more depth at A2. The reaction in Figure 10.8 is not tested at AS.

Propane-1,2,3-triol reacts with carboxylic acids to form triesters. When propane-1,2,3-triol reacts with naturally occurring fatty acids (e.g. palmitic acid, $C_{16}H_{32}O_2$; stearic acid, $C_{18}H_{36}O_2$; and oleic acid, $C_{18}H_{34}O_2$ — see Figure 10.9), triglycerides (lipids) are formed. Palmitic and stearic acids are saturated fatty acids; oleic acid is an unsaturated fatty acid because it contains a C=C double bond.

The 'tail' of a fatty acid molecule is a long hydrocarbon chain. When attached to propane-1,2,3-triol to form a triglyceride, the whole molecule is hydrophobic.

The terms saturated, monounsaturated and polyunsaturated refer to the number of C=C double bonds in the hydrocarbon tails of the fatty acids. A saturated fatty acid molecule has no double C=C bonds. The hydrocarbon chains in these fatty acids are fairly straight and pack closely together. This enables more surface interaction and creates more van der Waals forces, making the fats formed from saturated fatty acids solid at room temperature. Saturated fats are mostly obtained from animal sources.

A monounsaturated fatty acid molecule contains one double C=C bond.

A polyunsaturated fatty acid molecule has more than one C=C double bond in the hydrocarbon tail, causing bends or 'kinks' in the shape of the molecule. These kinks mean that molecules of unsaturated fats cannot pack closely together, which reduces the van der Waals forces. Hence, unsaturated fats are liquid at room temperature. They are obtained mostly from plant sources and are considered to be healthier than saturated fats.

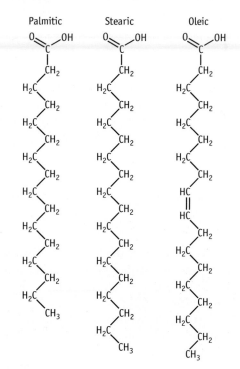

*Figure 10.9*

Since unsaturated fatty acids contain C=C double bonds, it is possible to have, *E/Z* isomers (Figure 10.10). Molecules in naturally occurring unsaturated vegetable oils are almost all *Z*. However, using oil for frying causes the conformation of some of the molecules to change to the *E*-form. If oil is used once, only a few of the bonds are converted. However, if oil is used continually, as in fast-food friers, more and more conformation changes occur and significant numbers of *trans* fatty acids are produced. This is of concern because fatty acids with the *E*-conformation can be carcinogenic.

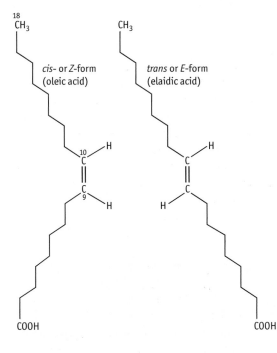

18
$CH_3$

cis- or Z-form
(oleic acid)

$CH_3$

trans or E-form
(elaidic acid)

10
C —H

C
9
H

H

H

COOH

COOH

**Figure 10.10**

## Halogenoalkanes

The initial attack by a nucleophile on a primary halogenoalkane involves the lone pair of electrons on a $Y^-$ ion forming a new bond with the carbon atom and, at the same time, the bonded pair of electrons in the C–X bond breaking away to form a lone pair of electrons on the resulting $X^-$. This initial attack is the slow step in the mechanism. It involves two reacting particles, one of which is a nucleophile (Y), and results in the substitution of X by Y. Hence, the mechanism is described as an $S_N2$ mechanism:

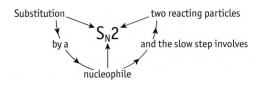

Substitution\
by a\
$S_N2$\
two reacting particles\
and the slow step involves\
nucleophile

The nucleophilic substitution of a tertiary halogeno-alkane occurs by the following mechanism, in which $(CH_3)_3CX$ is used as the example (see Figure 10.11):

■ Each of the alkyl groups, $CH_3$, pushes electrons along the σ-bond towards the central carbon atom. This causes the bonded pair in the C–X bond to break, forming a carbonium ion, $(CH_3)_3C^+$.

■ The carbonium ion, though unstable, is stabilised by the three inductive effects long enough for a nucleophile, $Y^-$, to attack the carbonium ion and form the product, $(CH_3)_3CY$ (Figure 10.11).

This mechanism also involves substitution (S) by a nucleophile ($_N$) but the initial slow step involves only a single molecule (1) — the tertiary halogenoalkane. Hence, it is described as an $S_N1$ mechanism.

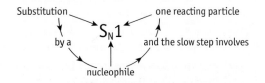

Substitution\
by a\
$S_N1$\
one reacting particle\
and the slow step involves\
nucleophile

Primary halogenoalkanes react by the $S_N2$ mechanism, which involves simultaneous bond breaking and bond making. Tertiary halogenoalkanes react by the $S_N1$ mechanism, which involves bond breaking followed by bond formation. Nucleophilic substitution of secondary halogenoalkanes can proceed by either mechanism or by a mixture of both.

## Elimination reactions

Changing the solvent from water to ethanol can alter the outcome of the hydrolysis of tertiary halogenoalkanes. For example, when 2-bromo-2-methylpropane reacts with hot ethanolic sodium hydroxide, it undergoes an elimination reaction, not a nucleophilic substitution reaction, and the product is 2-methylpropene. Secondary halogenoalkanes, and, with much greater difficulty, primary halogenoalkanes, can also undergo elimination reactions.

**Figure 10.11**

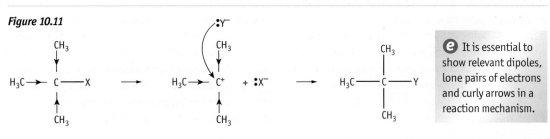

e It is essential to show relevant dipoles, lone pairs of electrons and curly arrows in a reaction mechanism.

Thus, when a halogenoalkane reacts with an alkali, depending on the conditions, two types of reaction are possible: nucleophilic substitution and elimination. Both may occur together to form a mixture of products (e.g. see Figure 10.12).

Primary halogenoalkanes usually undergo nucleophilic substitution; tertiary halogenoalkanes usually undergo elimination reactions. Secondary halogenoalkanes can react by either mechanism — the products depend on the conditions.

**Figure 10.12**

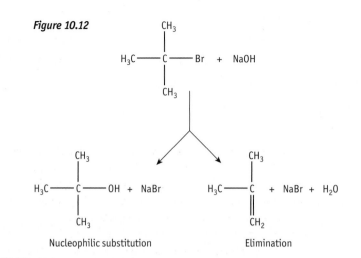

Nucleophilic substitution          Elimination

---

## Extension question: identify the isomers

There are four isomers, L, M, N and P, of molecular formula $C_4H_9X$, where X is a halogen.

a Draw skeletal formulae for the four isomers.

b Each isomer reacts with aqueous silver nitrate to produce a cream precipitate that dissolves in concentrated, but not dilute, aqueous ammonia. Identify X. Explain your answer.

c Write an ionic equation for the reaction of silver nitrate with X.

d Isomer L can be hydrolysed to produce a compound, $C_4H_{10}O$, that does *not* react when warmed with a mixture of potassium dichromate(VI) and sulphuric acid. Identify isomer L. Explain your answer.

e Isomer M can also be hydrolysed. The compound obtained from this reaction *does* react with a mixture of potassium dichromate(VI) and sulphuric acid. The organic product obtained from this reaction is neutral. Identify isomer M. Explain your answer and write equations for the reactions described.

f The isomers N and P react in a similar way. Each produces an alkane when the following reactions are carried out:

(i) Compounds N and P are hydrolysed to produce $C_4H_{10}O$.

(ii) $C_4H_{10}O$ is then dehydrated to $C_4H_8$.

(iii) $C_4H_8$ is reacted with hydrogen using a nickel catalyst.

The alkane formed from compound N has a lower boiling point than that formed from compound P. Identify N and P. Explain your answers.

g Consider reactions (i)–(iii) in part f above. Place them in order of decreasing atom economy. Calculations are not necessary, but you must justify your answer.

h The mass spectrum of isomer P is shown below.

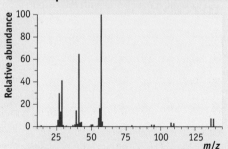

(i) Calculate the relative molecular mass of the isomers.

(ii) There are two peaks of equal height at $m/z$ 136 and 138. Suggest why these two peaks occur.

(iii) The mass spectrum also has a prominent line at $m/z$ 57. Suggest the source of this peak.

(iv) There is also a peak at $m/z$ 29. Explain how this peak would enable isomer P to be distinguished from isomer L.

# Energy

## Chapter 11

Why some chemicals react together and others do not is a fundamental question central of the study of chemistry. No fully comprehensive theory has been formulated to answer this difficult question. One way that chemists have tried to address this issue is through the understanding of atomic structure. Another approach is to analyse the energy changes that occur in reactions and then use this analysis to predict what might happen. Although simplistic, it is often true that reactions in which the products have lower energy than the reactants *do* take place, whereas reactions in which the products have higher energy do not. An important point, however, is that the rate of conversion from reactants to products may be extremely slow. Thus, a reaction might potentially proceed, but progress may not be apparent.

In the first section of this chapter, the methods used to determine the energy changes that take place when reactions occur are introduced. A further section covers issues that affect how readily reactions take place. The final section considers cases where reactions do not go to completion and reactants and products are both present at the end of the reaction.

## Energetics

First, it is reasonable to ask what is meant by 'energy'. It could be interpreted as 'vigour'. Indeed, physical vigour is one manifestation of energy — **kinetic energy**, energy due to movement. Other forms of energy are heat, light and sound. However, energy is more subtle than this. **Potential energy** is contained by an object on account of its structure and/or position. This is energy that can be thought of as waiting to be released. Energy contained within chemicals is an important driving force in the way that they behave. Atoms, bonds and structure all contribute to this energy.

> Energy can be neither created nor destroyed but can be converted from one form to another.

In this section, the chemical energy stored in molecules and the ways in which it can be converted into other forms of energy are investigated. In chemistry, energy is measured in joules, J, or kilojoules, kJ.

### Enthalpy change

Almost all chemical reactions involve a change in energy. The exchange of energy between a reaction mixture and its surroundings is the **enthalpy change**.

The enthalpy change is measured at constant pressure. It is represented by the symbol $\Delta H$. The units are normally $kJ\,mol^{-1}$.

The symbol $\Delta$ is used to indicate 'change in':
- $\Delta T$ = change in temperature
- $\Delta V$ = change in volume
- $\Delta P$ = change in pressure
- $\Delta H$ = change in enthalpy
  
  **$\Delta H$ = enthalpy of products – enthalpy of reactants**

When an **exothermic reaction** takes place, energy is transferred from the reaction mixture to the surroundings. Chemical energy is released by the reactants and the temperature of the surroundings increases. This happens because the chemical energy of the reactants is greater than the chemical energy of the products and, since energy cannot be created or destroyed, the difference is transferred to the surroundings.

◁ In an **exothermic** reaction, $\Delta H$ is **negative**.

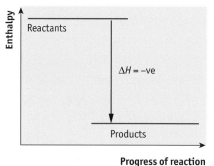

*Figure 11.1 Enthalpy-level diagram for an exothermic reaction*

*Fireworks involve exothermic reactions*

When an **endothermic reaction** takes place, heat energy is transferred from the surroundings (or from an external source) to the reaction mixture. Endothermic reactions require energy — the temperature of the surroundings decreases. If the reaction mixture gains energy from its surroundings, the reaction is endothermic. In endothermic reactions, the products have more chemical energy than the reactants, so energy from the surroundings is required and is converted into chemical energy.

In an **endothermic** reaction, $\Delta H$ is **positive**.

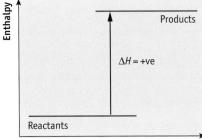

*Figure 11.2 Enthalpy-level diagram for an endothermic reaction*

## Enthalpy-profile diagrams

Enthalpy changes can be represented by simple enthalpy-profile diagrams.

For an exothermic reaction, the enthalpy-profile diagram (Figure 11.3) shows the products at a lower energy than the reactants. The difference in enthalpy is $\Delta H$.

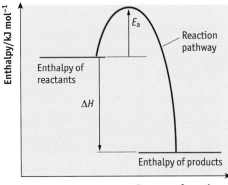

*Figure 11.3 Enthalpy-profile diagram for an exothermic reaction*

The reaction pathway in Figure 11.3 shows that, initially, energy is taken from the surroundings and then energy is released to the surroundings. Reactions need an input of energy to get them started. This is called the **activation energy**, $E_a$.

Activation energy is the minimum energy needed for colliding particles to react.

In a chemical reaction, bonds are broken and new bonds are formed. Breaking a bond is endothermic and, therefore, requires energy. This energy requirement contributes to the activation energy of a reaction. Enthalpy-profile diagrams show both the enthalpy change of reaction, $\Delta H$, and the activation energy, $E_a$.

Activation energy is covered in more detail in the section on reaction rates (page 213).

### Enthalpy change in an exothermic reaction

All combustion reactions are exothermic. The chemical energy stored in the molecules supplies heat to the surroundings. This results in an increase in temperature of the surroundings.

The enthalpy profile in Figure 11.4 illustrates the combustion of methane.

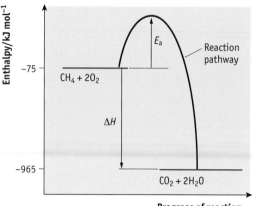

*Figure 11.4*
*Enthalpy-profile diagram for the combustion of methane.*

If the enthalpies of the reactants and products are known, the following equation can be used to calculate the **enthalpy of combustion**:

$\Delta H =$ enthalpy of products − enthalpy of reactants

For example, for methane this is:

$\Delta H = (-965) - (-75) = -965 + 75 = -890 \, kJ \, mol^{-1}$

Notice that the enthalpies of both reactants and products are negative.

### Enthalpy change in an endothermic reaction

For an **endothermic reaction**, the enthalpy-profile diagram shows the products at a higher energy level than the reactants (Figure 11.5). The difference in the enthalpy is $\Delta H$.

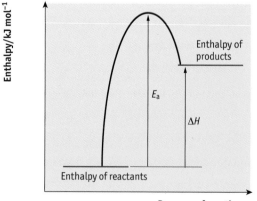

*Figure 11.5*
*Enthalpy-profile diagram for an endothermic reaction*

Endothermic reactions also require a minimum amount of energy, $E_a$, to initiate the reaction. However, in this case, the activation energy, $E_a$, is greater than the change in enthalpy, $\Delta H$.

Thermal decomposition reactions are usually endothermic and, therefore, require energy from the surroundings. The enthalpy profile for the thermal decomposition of calcium carbonate is shown in Figure 11.6.

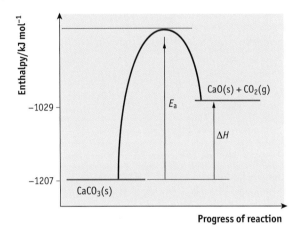

**Figure 11.6**
*Enthalpy-profile diagram for thermal decomposition of calcium carbonate*

The enthalpy of decomposition of calcium carbonate can be calculated using the equation:

$\Delta H$ = enthalpy of products − enthalpy of reactants

$\Delta H = -1029 - (-1207) = -1029 + 1207 = +178\,kJ\,mol^{-1}$

**Enthalpy change in a reversible reaction**

Many reactions are reversible. This means that the reaction takes place simultaneously in both the forward and reverse directions. The reaction is exothermic in one direction and endothermic in the other direction.

Ammonia, $NH_3$, is formed by the reaction between nitrogen and hydrogen (see page 222). The reaction is reversible (Figure 11.7).

$N_2(g) + 3H_2(g) \rightleftharpoons 2NH_3(g)$

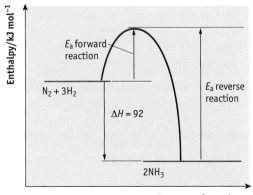

**Figure 11.7**
*Enthalpy-profile diagram for the reversible formation of ammonia*

ⓔ In calculations of this kind, it is easy to make an error in the use of minus signs, so care is needed.

The numerical value of the enthalpy of reaction, $\Delta H$, is the same for the forward and the reverse reactions. However, the sign is different:

- The forward reaction is exothermic; $\Delta H = -92\,\text{kJ}\,\text{mol}^{-1}$.
- The reverse reaction is endothermic; $\Delta H = +92\,\text{kJ}\,\text{mol}^{-1}$.

The activation energies for the forward and the reverse reactions are different:

- $E_a$ for the forward reaction is $+68\,\text{kJ}\,\text{mol}^{-1}$.
- $E_a$ for the reverse reaction is $+92 + 68 = +160\,\text{kJ}\,\text{mol}^{-1}$.

## Standard enthalpy changes

Enthalpy changes are measured experimentally at various temperatures and pressures. However, these are converted to values that would apply under **standard conditions**:

- standard temperature = 298 K (25°C)
- standard pressure = 101 kPa

Standard temperature and pressure are abbreviated to s.t.p.

The superscripted symbol, $\ominus$, is used with $\Delta H$ to denote the standard enthalpy change, i.e. $\Delta H^{\ominus}$.

> For any reaction, the standard enthalpy change of reaction, $\Delta H^{\ominus}_r$, is the enthalpy change when the number of moles of the substances in the equation *as written* react under standard conditions of 298 K and 101 kPa.

Exam questions often ask for definitions of specific enthalpy changes and it is advisable to learn the following:

> The standard enthalpy change of formation, $\Delta H^{\ominus}_f$, is the enthalpy change when 1 mol of a substance is formed from its elements, in their natural state, under standard conditions of 298 K and 101 kPa.

> The standard enthalpy change of combustion, $\Delta H^{\ominus}_c$, is the enthalpy change when 1 mol of a substance is burnt completely, in an excess of oxygen, under standard conditions of 298 K and 101 kPa.

> The standard enthalpy change of neutralisation, $\Delta H^{\ominus}_{neut}$, is the enthalpy change when 1 mol of an acid (or a base) is neutralised by a base (or an acid) under standard conditions of 298 K and 101 kPa.

### Equations representing enthalpy change

You may have to show your understanding by writing equations illustrating the standard enthalpy changes of formation, combustion and neutralisation.

### Standard of enthalpy of formation

An equation relating to the standard enthalpy of formation of a substance must reflect the standard definition. Therefore, it must:

- show the reactants as elements
- produce 1 mol of the substance, even if this means that there are fractions in the equation
- include state symbols

The equation for the standard enthalpy of formation of ethane is:

$$2C(s) + 3H_2(g) \rightarrow C_2H_6(g)$$

The equation for the standard enthalpy of formation of ethanol is:

$$2C(s) + 3H_2(g) + \tfrac{1}{2}O_2(g) \rightarrow CH_3CH_2OH(l)$$

The equation for the standard enthalpy of formation of ethanamide is:

$$2C(s) + 2\tfrac{1}{2}H_2(g) + \tfrac{1}{2}O_2(g) + \tfrac{1}{2}N_2(g) \rightarrow CH_3CONH_2(s)$$

*Standard enthalpy of combustion*
An equation relating to the standard enthalpy of combustion of a substance must reflect the standard definition. Therefore, it must:
- show 1 mol of the substance reacting completely with excess oxygen, even if this means having fractions in the equation
- include state symbols

The products of combustion are usually carbon dioxide and water.

The equation for the standard enthalpy of combustion of ethane is:

$$C_2H_6(g) + 3\tfrac{1}{2}O_2(g) \rightarrow 2CO_2(g) + 3H_2O(l)$$

The equation for the standard enthalpy of combustion of ethanol is:

$$CH_3CH_2OH(l) + 3O_2(g) \rightarrow 2CO_2(g) + 3H_2O(l)$$

The equation for the enthalpy of combustion of ethanamide is more complex, because, under standard conditions, the nitrogen in the ethanamide does not combine with oxygen. The equation is also difficult to balance. However, the principles are the same — the equation must have 1 mol of the reactant and be balanced, even though this involves odd-looking fractions:

$$CH_3CONH_2(s) + 2\tfrac{3}{4}O_2 \rightarrow 2CO_2(g) + 2\tfrac{1}{2}H_2O(l) + \tfrac{1}{2}N_2(g)$$

*Standard enthalpy of neutralisation*
An equation relating to the standard enthalpy of neutralisation of a substance must reflect the standard definition. It must:
- show 1 mol of the substance (acid or base) reacting with the named base or acid, even if this means having fractions in the equation
- include state symbols

Usually the products are a salt and water.

The equation for the standard enthalpy of neutralisation of dilute hydrochloric acid with aqueous sodium hydroxide is:

$$HCl(aq) + NaOH(aq) \rightarrow NaCl(aq) + H_2O(l)$$

An ionic equation can also be used to show the standard enthalpy of neutralisation:

$$H^+(aq) + OH^-(aq) \rightarrow H_2O(l)$$

## Measuring enthalpy changes experimentally

It is easy to measure the temperature changes for reactions that take place in solution and for combustion reactions.

### Reactions in solution

The standard enthalpy change, $\Delta H_r^\ominus$, for reactions that take place in solution can usually be measured directly using the apparatus shown in Figure 11.8. The result obtained is approximate because of heat loss to the surroundings.

**Figure 11.8** *Apparatus to measure the enthalpy of reaction*

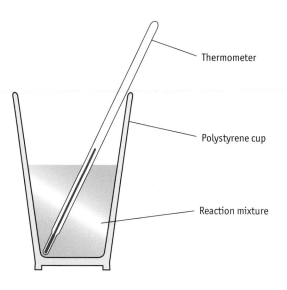

The energy transfer for the reaction mixture, $Q$, can be calculated using the equation:

$$Q = mc\Delta T$$

- $m$ is the mass of the reaction mixture
- $c$ is the specific heat capacity of the reaction mixture
- $\Delta T$ is the change in temperature

The **specific heat capacity** of a substance is the heat required to increase the temperature of 1.0 g of the substance by 1°C (1 K).

The solvent used for most reactions is water. The specific heat capacity of water is $4.20\,J\,g^{-1}\,K^{-1}$ or $4.20\,kJ\,kg^{-1}\,K^{-1}$.

The enthalpy change for the reaction mixture, $Q$, has a value in either joules (J) or kilojoules (kJ), depending on the units of specific heat capacity. It is usual to adjust this value so that $\Delta H$ can be quoted for 1 mol of reactant with the units of $kJ\,mol^{-1}$. This may require mole calculations (page 40).

The temperature units can be either degrees Celsius or Kelvin.

The standard enthalpy change for the reaction is calculated by dividing the energy transferred by the number of moles, $n$, of reactant consumed:

$$\Delta H_r^\ominus = \frac{Q}{n} = \frac{mc\Delta T}{n}$$

> **Worked example**
>
> When $50.0\,cm^3$ of $2.00\,mol\,dm^{-3}$ hydrochloric acid is mixed with $50.0\,cm^3$ of $2.00\,mol\,dm^{-3}$ sodium hydroxide solution, the temperature increases by $13.7°C$.
>
> Calculate the standard enthalpy change of neutralisation of hydrochloric acid. Assume that the specific heat capacity, $c = 4.20\,J\,g^{-1}\,K^{-1}$ and that the density of both the hydrochloric acid and sodium hydroxide solution is $1.00\,g\,cm^{-3}$.
>
> **Answer**
> **Step 1:** calculate the energy transferred in the reaction by using:
> $Q = mc\Delta T$
> $m$ = mass of the two solutions = $50.0 + 50.0 = 100.0\,g$
> $Q = mc\Delta T = 100.0 \times 4.20 \times 13.7 = 5754\,J = 5.754\,kJ$
> **Step 2:** convert the answer to $kJ\,mol^{-1}$ by dividing by the number of moles used
>
> amount, $n$ in moles of HCl = $cV = 2.00 \times \dfrac{50.0}{1000} = 0.100\,mol$
>
> $\Delta H_{neut} = \dfrac{Q}{n} = \dfrac{5.754}{0.1} = 57.54 = 57.5\,kJ\,mol^{-1}$ (3 s.f.)
>
> Remember that because the temperature rises, this is an exothermic reaction. Therefore, $\Delta H_{neut} = -57.5\,kJ\,mol^{-1}$.

In the worked example above, the volumes and concentrations of the hydrochloric acid and sodium hydroxide solution were such that the acid would react completely. Two other situations might arise in an experiment of this type:

- The concentration (or volume) of the sodium hydroxide solution might be such that the amount (in moles) of sodium hydroxide is greater than the amount (in moles) of hydrochloric acid. The hydrochloric acid would be neutralised completely, but some unreacted sodium hydroxide would remain. This would not affect the value of $\Delta H$.
- The concentration (or volume) of sodium hydroxide might be insufficient to ensure that all the hydrochloric acid is neutralised. This would affect the value of $\Delta H$ because some unreacted hydrochloric acid would remain. In this case, the reaction is being controlled by the amount of sodium hydroxide. The value of $\Delta H$ calculated must be based on the amount (in moles) of sodium hydroxide and the amount (in moles) of hydrochloric acid that would, therefore, have been neutralised.

**e** Be careful not to round any numbers during the calculation.

**e** The final answer is quoted to three significant figures because the data are given to three significant figures.

**e** In neutralisation reactions, an excess of the neutralising reagent should be used.

### Combustion

The standard enthalpy of combustion, $\Delta H_c^{\ominus}$, for a volatile liquid (e.g. ethanol) can be measured directly using the apparatus shown in Figure 11.9. The result will be approximate because of heat loss.

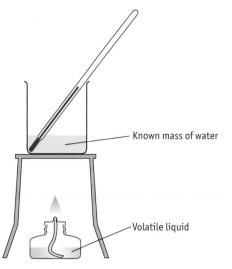

Known mass of water

Volatile liquid

*Figure 11.9*
*Apparatus used to determine the enthalpy of combustion of a volatile liquid, such as ethanol*

---

**Worked example**

When 0.230 g ethanol, $CH_3CH_2OH$, is burned, it raises the temperature of 200 cm³ of water by 7.80°C. Calculate the standard enthalpy change of combustion of ethanol. Assume that the specific heat capacity, $c$, is $4.20\,J\,g^{-1}\,K^{-1}$ and that the density of water is $1.00\,g\,cm^{-3}$.

**Answer**

**Step 1:** calculate the energy transferred in the reaction by using:

$Q = mc\Delta T$

$m$ = mass of water = 200 g

$Q = mc\Delta T = 200 \times 4.20 \times 7.80 = 6552\,J = 6.552\,kJ$

**Step 2:** convert the answer to $kJ\,mol^{-1}$ by dividing by the number of moles used

$$\text{amount, } n \text{ in moles of ethanol} = \frac{\text{mass of ethanol}}{M_r \text{ of ethanol}} = \frac{0.230}{46} = 0.00500\,mol$$

$$\Delta H = \frac{Q}{n} = \frac{6.552}{0.00500} = 1310\,kJ\,mol^{-1}$$

---

## Hess's law

The enthalpy changes for combustion reactions and for reactions that take place in solution can be measured experimentally. However, for energetic or kinetic reasons, this is not the case for many chemical reactions. If the activation energy is particularly high or the reaction rate is very slow (these are usually related), it is unlikely that a reaction will take place readily. In these circumstances, enthalpy change cannot be measured directly.

Enthalpies that cannot be measured directly are calculated by using energy cycles. Energy cannot be created or destroyed and, therefore, the enthalpy change for a reaction is independent of the route taken. This was first proposed by Germain Henri Hess and is known as Hess's law.

> Hess's law states that, if a reaction can take place by more than one route, the enthalpy change for the reaction is the same irrespective of the route taken, provided that the initial and final conditions are the same.

◄ Hess's law may seem an obvious variation of the law of conservation of energy. However, Hess stated his law in 1840, before the law of conservation of energy was generally accepted.

When applying Hess's law it is helpful to construct an enthalpy cycle:

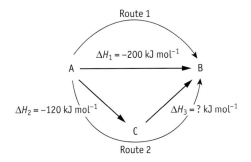

The enthalpy change for route 1 is equal to the enthalpy change for route 2.

$$\Delta H_1 = \Delta H_2 + \Delta H_3$$

Therefore, $\Delta H_3 = \Delta H_1 - \Delta H_2$
$$= -200 - (-120)$$
$$= -200 + 120$$
$$= -80 \, \text{kJ} \, \text{mol}^{-1}$$

### Enthalpy of formation of carbon monoxide

The usefulness of Hess's law can be illustrated by the combustion of carbon. Carbon burns to form either $CO(g)$ or $CO_2(g)$. It is impossible to measure the enthalpy of formation of $CO(g)$ directly, but it can be calculated using Hess's law.

The enthalpy change for the following reactions can be measured experimentally:

$$C(s) + O_2(g) \rightarrow CO_2(g) \qquad \Delta H_c = -394 \, \text{k} \, \text{mol}^{-1}$$
$$CO(g) + \tfrac{1}{2}O_2(g) \rightarrow CO_2(g) \qquad \Delta H_c = -284 \, \text{kJ} \, \text{mol}^{-1}$$

It is not possible to measure directly the enthalpy change for the conversion of carbon into carbon monoxide:

$$C(s) + \tfrac{1}{2}O_2(g) \rightarrow CO(g)$$

However, this enthalpy change can be calculated by constructing an enthalpy cycle.

**Step 1:** start with the enthalpy change that has to be calculated. Call it $\Delta H_1$. Write an equation for the reaction. This is the top line of the cycle:

$$\Delta H_1$$
$$C(s) + \tfrac{1}{2}O_2(g) \rightarrow CO(g)$$

**Step 2:** construct an enthalpy cycle with two alternative routes:

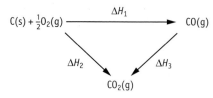

**Step 3:** apply Hess's law to the cycle.

The simplest way to apply Hess's law is to look at the direction of the arrows. Route 1 has arrows that point in the clockwise direction; route 2 has an arrow that points anti-clockwise:

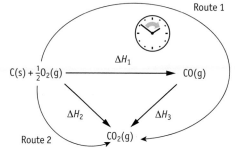

route 1 = route 2

$\Delta H_1 + \Delta H_3 = \Delta H_2$

Therefore, $\Delta H_1 = \Delta H_2 - \Delta H_3$

$$= -394 - (-284)$$
$$= -394 + 284$$
$$= -110 \, \text{kJ mol}^{-1}$$

The enthalpy of formation of carbon monoxide is $-110 \, \text{kJ mol}^{-1}$.

## Construction of enthalpy cycles

There are two possible enthalpy cycles that you may be asked to construct.

### Enthalpy of formation from enthalpies of combustion

In order to calculate the enthalpy of formation using enthalpies of combustion, it is best to construct a cycle with the combustion products at the bottom. The arrows always then point downwards to the combustion products.

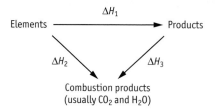

Applying Hess's law to the cycle:

$\Delta H_1 + \Delta H_3 = \Delta H_2$

Therefore, $\Delta H_1 = \Delta H_2 - \Delta H_3$

### Enthalpy of combustion from enthalpies of formation

The alternative calculation involves calculating the enthalpy of combustion from enthalpy of formation data. The elements are best shown at the bottom of the triangle; the arrows then point upwards.

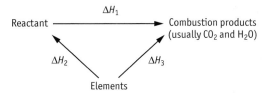

Applying Hess's law to the cycle:

$$\Delta H_1 + \Delta H_2 = \Delta H_3$$

Therefore, $\Delta H_1 = \Delta H_3 - \Delta H_2$

---

**Worked example 1**

Use the data in the table below to calculate the standard enthalpy of formation for methane.

| Substance | Formula | $\Delta H_c^{\ominus}/\text{kJ mol}^{-1}$ |
|-----------|---------|------------------------|
| Carbon | $C(s)$ | −394 |
| Hydrogen | $H_2(g)$ | −286 |
| Methane | $CH_4(g)$ | −890 |

**Answer**

**Step 1:** write the equation for the required enthalpy of formation:

$$C(s) + 2H_2(g) \rightarrow CH_4(g)$$

**Step 2:** construct the enthalpy cycle, writing the combustion products at the bottom. Both arrows point downwards:

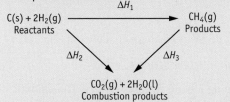

**Step 3:** apply Hess's law using the clockwise equals anti-clockwise rule.

Using Hess's law, $\Delta H_1 + \Delta H_3 = \Delta H_2$

Therefore, $\Delta H_1 = \Delta H_2 - \Delta H_3$

$$\Delta H_2 = \Delta H_c(C) + 2 \times (\Delta H_c(H_2))$$
$$= -394 + 2(-286)$$
$$= -966 \text{ kJ mol}^{-1}$$

$$\Delta H_3 = \Delta H_c(CH_4) = -890 \text{ kJ mol}^{-1}$$

$$\Delta H_1 = -966 - (-890) = -76 \text{ kJ mol}^{-1}$$

Therefore, the standard enthalpy of formation of methane, $\Delta H_f^{\ominus}(CH_4)$, is −76 kJ mol⁻¹.

**e** Remember the clock.

## Worked example 2

Use the data in the table below to calculate the standard enthalpy of combustion for ethane.

| Substance | Formula | $\Delta H_f^{\ominus}/kJ\,mol^{-1}$ |
|---|---|---|
| Ethane | $C_2H_6(g)$ | −85 |
| Carbon dioxide | $CO_2(g)$ | −394 |
| Water | $H_2O(l)$ | −286 |

**Answer**

**Step 1:** write the equation for the required enthalpy of combustion:

$$C_2H_6(g) + 3\tfrac{1}{2}O_2(g) \rightarrow 2CO_2(g) + 3H_2O(l)$$

**Step 2:** construct the enthalpy cycle, writing the combustion products at the bottom. Both arrows point upwards:

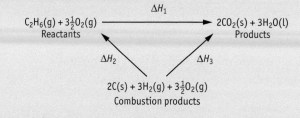

Using Hess's law, $\Delta H_1 + \Delta H_2 = \Delta H_3$

Therefore, $\Delta H_1 = \Delta H_3 - \Delta H_2$

$\Delta H_2$ for formation of $C_2H_6(g) = -85\,kJ\,mol^{-1}$

$$\Delta H_3 = 2 \times \Delta H_f(CO_2) + 3 \times \Delta H_f(H_2O)$$
$$= 2(-394) + 3(-286)$$
$$= -1646\,kJ\,mol^{-1}$$

$\Delta H_1 = -1646 - (-85) = -1561\,kJ\,mol^{-1}$

Therefore, the standard enthalpy of combustion of ethane, $\Delta H_c^{\ominus}(C_2H_6)$, is $-1561\,kJ\,mol^{-1}$.

# Bond enthalpy

## Average bond enthalpy

Breaking a bond requires energy; forming a bond releases energy. The amount of energy required to break a bond is the same as the energy released when the bond is formed.

**Bond (dissociation) enthalpy** is the energy required to break a bond. It is always endothermic ($+\Delta H$). Breaking a bond is equivalent to homolytic fission — it produces two neutral particles (radicals) in the gaseous state:

$$H-Cl(g) \rightarrow H{\bullet}\,(g) + Cl{\bullet}\,(g)$$

Bond enthalpy is the energy required to break the same bond in all the molecules of 1 mol of a gas and to separate the resulting gaseous (neutral) particles/atoms/radicals so that they exert no forces on each other.

It is best reinforced by a simple equation such as:

$$Cl-Cl(g) \rightarrow Cl{\bullet}\,(g) + Cl{\bullet}\,(g)$$

In general:
$$X–Y(g) \rightarrow X\bullet(g) + Y\bullet(g)$$

Bond enthalpies are average (mean) values and do not take into account the specific chemical environment.

In a molecule such as water, the O–H bonds can be broken successively. The energy required is different for each bond:

$$\Delta H = +496 \text{ kJ mol}^{-1}$$

$$\Delta H = +432 \text{ kJ mol}^{-1}$$

The O–H bond enthalpy in water is quoted as $\dfrac{(496 + 432)}{2} = +464 \text{ kJ mol}^{-1}$.

This is the average (mean) bond enthalpy.

A similar situation applies to the C–H bond in methane. There are four C–H bonds that can be broken successively. The energy required is different for each bond, as shown in Table 11.1.

| Successive bond breakage | Bond enthalpy, $\Delta H$/kJ mol$^{-1}$ |
| --- | --- |
| $CH_4(g) \rightarrow CH_3(g) + H(g)$ | +425 |
| $CH_3(g) \rightarrow CH_2(g) + H(g)$ | +470 |
| $CH_2(g) \rightarrow CH(g) + H(g)$ | +416 |
| $CH(g) \rightarrow C(g) + H(g)$ | +335 |

*Table 11.1 Bond enthalpies of the C–H bonds in methane*

The average bond enthalpy for a C–H bond in methane is:
$$\frac{+425 + 470 + 416 + 335}{4} = \frac{+1646}{4} = +411.5 \text{ kJ mol}^{-1}$$

The C–H bond enthalpy quoted in most books is $+413 \text{ kJ mol}^{-1}$. This is slightly different from the average bond enthalpy of the C–H bonds in methane and is the average C–H bond enthalpy determined for a *range* of organic compounds.

Some average bond enthalpies are shown in Table 11.2.

| Bond | $\Delta H$/kJ mol$^{-1}$ |
| --- | --- |
| C–H | +413 |
| C=O | +805 |
| O=O | +498 |
| O–H | +464 |
| C–N | +286 |
| C=C | +612 |
| N≡N | +945 |
| N–H | +391 |
| C–C | +347 |
| H–H | +436 |

*Table 11.2 Average bond enthalpies*

## Calculations involving average bond enthalpy

Calculations are straightforward. For a reaction to take place, existing bonds have first to be broken and new bonds have to be formed.

For a simple reaction involving gaseous molecules:

$$\Delta H = \Sigma \text{ bond enthalpies of reactants} - \Sigma \text{ bond enthalpies of products}$$

◀ Bond breaking is endothermic. Bond formation is exothermic.

All calculations use average bond enthalpies. As shown in the worked examples, it is useful to draw out the reaction using displayed formulae, so that all the bonds broken and formed can be seen easily.

---

### Worked example 1

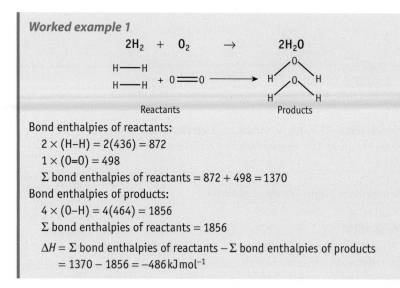

$$2H_2 \quad + \quad O_2 \quad \rightarrow \quad 2H_2O$$

Reactants         Products

Bond enthalpies of reactants:

$2 \times (\text{H–H}) = 2(436) = 872$

$1 \times (\text{O=O}) = 498$

$\Sigma$ bond enthalpies of reactants $= 872 + 498 = 1370$

Bond enthalpies of products:

$4 \times (\text{O–H}) = 4(464) = 1856$

$\Sigma$ bond enthalpies of reactants $= 1856$

$\Delta H = \Sigma$ bond enthalpies of reactants $- \Sigma$ bond enthalpies of products

$= 1370 - 1856 = -486\,\text{kJ}\,\text{mol}^{-1}$

---

### Worked example 2

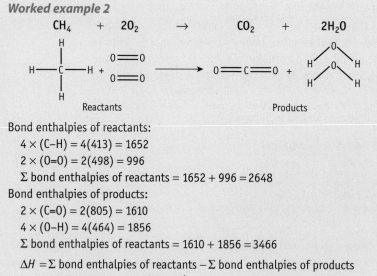

$$CH_4 \quad + \quad 2O_2 \quad \rightarrow \quad CO_2 \quad + \quad 2H_2O$$

Reactants         Products

Bond enthalpies of reactants:

$4 \times (\text{C–H}) = 4(413) = 1652$

$2 \times (\text{O=O}) = 2(498) = 996$

$\Sigma$ bond enthalpies of reactants $= 1652 + 996 = 2648$

Bond enthalpies of products:

$2 \times (\text{C=O}) = 2(805) = 1610$

$4 \times (\text{O–H}) = 4(464) = 1856$

$\Sigma$ bond enthalpies of reactants $= 1610 + 1856 = 3466$

$\Delta H = \Sigma$ bond enthalpies of reactants $- \Sigma$ bond enthalpies of products

$= 2648 - 3466 = -818\,\text{kJ}\,\text{mol}^{-1}$

---

The accepted enthalpy change for the combustion of methane is $-890\,kJ\,mol^{-1}$, which differs substantially from the value $-818\,kJ\,mol^{-1}$ calculated in the worked example above. This is largely explained by using average bond enthalpies for the C–H, C=O and O–H bonds in the calculation, rather than specific bond enthalpies.

## Questions

**1** Explain, with the aid of enthalpy-profile diagrams, what is meant by an exothermic reaction and an endothermic reaction. On each diagram, label both $\Delta H$ and $E_a$.

**2** When a mixture of hydrogen gas and oxygen gas is ignited by a spark, water is produced:

$$H_2(g) + \tfrac{1}{2}O_2(g) \rightarrow H_2O(l) \quad \Delta H = -285.8\,kJ\,mol^{-1}$$
$$H_2(g) + \tfrac{1}{2}O_2(g) \rightarrow H_2O(g) \quad \Delta H = -241.8\,kJ\,mol^{-1}$$

**a** On the same diagram, draw the enthalpy profile for the formation of both water and steam.

**b** Use the enthalpy-profile diagram to deduce the enthalpy change for the conversion of water into steam:

$$H_2O(l) \rightarrow H_2O(g)$$

**3** State the standard condition of:

**a** temperature

**b** pressure

**4 a** Define standard enthalpy of formation.

**b** Write an equation to illustrate the standard enthalpy of formation of propanal, $CH_3CH_2CHO(l)$.

**5 a** Define standard enthalpy of combustion.

**b** Write an equation to illustrate the standard enthalpy of combustion of propanone, $CH_3COCH_3(l)$.

**6** When $2.0\,cm^3$ of concentrated sulphuric acid is added to $48.0\,cm^3$ of water, the temperature of the mixture rises by 6.7°C. Calculate the amount of heat evolved. Assume that the specific heat capacity of the mixture is $4.2\,J\,g^{-1}\,K^{-1}$.

**7** When $25.0\,cm^3$ of nitric acid of concentration $1.00\,mol\,dm^{-3}$ is mixed with $50\,cm^3$ of $0.50\,mol\,dm^{-3}$ sodium hydroxide solution, the temperature increases by 4.6°C. Calculate the standard enthalpy change of neutralisation of nitric acid. Assume that the specific heat capacity, $c$, of the mixture is $4.20\,J\,g^{-1}\,K^{-1}$ and that the density of both the nitric acid and sodium hydroxide solutions is $1.00\,g\,cm^{-3}$.

**8** Use the data in the table below to calculate the standard enthalpy change of formation for propane.

| Substance | Formula | $\Delta H_f^{\ominus}/kJ\,mol^{-1}$ |
| --- | --- | --- |
| Carbon | C(s) | −394 |
| Hydrogen | $H_2(g)$ | −286 |
| Propane | $C_3H_8(g)$ | −2219 |

**9** Use the data in the table below to calculate the standard enthalpy change of combustion for ethanol.

| Substance | Formula | $\Delta H^{\ominus}_f$/kJ mol$^{-1}$ |
|---|---|---|
| Ethanol | $C_2H_5OH(g)$ | −277 |
| Carbon dioxide | $CO_2(g)$ | −394 |
| Water | $H_2O(l)$ | −286 |

**10 a** Explain, with the aid of an equation, what is meant by bond enthalpy.
   **b** Some bond enthalpies are given in the following table:

| Bond | Bond enthalpy/kJ mol$^{-1}$ |
|---|---|
| H–H | 436 |
| Cl–Cl | 327 |
| H–Cl | 432 |

Use the data in the table to calculate the enthalpy change of reaction, $\Delta H$, for:

$$H_2(g) + Cl_2(g) \rightarrow 2HCl(g)$$

**11** A student is given an unknown alcohol that has a standard enthalpy of combustion of −726 kJ mol$^{-1}$.

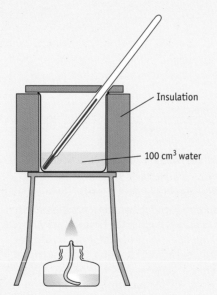

Using the above apparatus, the student obtains the following results:
- mass of water = 100 g
- initial temperature = 18.69°C
- final temperature = 25.66°C
- mass of spirit burner + alcohol at the start = 117.416 g
- mass of spirit burner + alcohol at the end = 117.288 g

The heat capacity, $c$, of the apparatus is 4.18 J g$^{-1}$ K$^{-1}$. Calculate the relative molecular mass of the alcohol and, hence, identify it.

# Reaction rates

Experimental observations show that the rate of reaction is influenced by temperature, concentration, the use of a catalyst and, for gaseous reactions, pressure.

The collision theory of reactivity helps to explain these observations. A reaction cannot take place unless a collision occurs between the reacting particles. The rate of a reaction can be increased by:

- increasing the concentration
- increasing the temperature
- increasing the pressure in a gaseous reaction

These factors increase the chance of a collision occurring. However, increasing the temperature also increases the energy of the collision, which is a more important consideration.

## Effect of concentration

A useful analogy is to imagine your first driving lesson. The one thing you want to avoid is a collision. It follows that your first lesson is likely to be early on Sunday morning on a quiet road, rather than at 5.00 p.m. on Friday evening in the city centre. It is obvious that the high concentration of cars at rush hour increases the chance of a collision. The same is true for a chemical reaction. Increasing the concentration increases the chance of a collision. The more collisions there are, the faster the reaction.

For a gaseous reaction, increasing the pressure has the same effect as increasing the concentration. When gases react, they react faster at high pressure because there is an increased chance of a collision.

Not all collisions lead to a successful reaction. For reaction to occur:

- the colliding particles must be correctly orientated
- the energy of a collision between reacting particles must exceed the **activation energy, $E_a$,** which is the minimum energy required for a reaction to occur

### Importance of correct orientation

The need for colliding particles to be correctly orientated is illustrated by the reaction between iodomethane, $CH_3I$, and hydroxy ions, $OH^-$ (Figure 11.10). In iodomethane, the C–I bond is polar. The carbon has a $\delta+$ charge; the iodine has a $\delta-$ charge. If a negatively charged hydroxy ion approaches iodomethane towards the $I^{\delta-}$ there is mutual repulsion and reaction does not occur. If the approach is towards the $C^{\delta+}$, the collision leads to a reaction.

*Figure 11.10 Effect of orientation on reaction outcome*

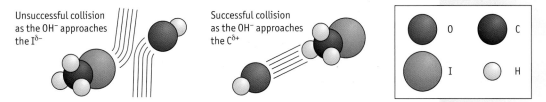

Unsuccessful collision as the $OH^-$ approaches the $I^{\delta-}$

Successful collision as the $OH^-$ approaches the $C^{\delta+}$

O   C   I   H

## Effect of temperature

Raising the temperature increases the energy of the particles — they move faster and collide more often. However, this increased frequency of collisions does not explain fully the effect that raising the temperature has on the rate of reaction.

When a collision occurs, the particles exchange (gain or lose) energy, even if a reaction does not occur. It follows that, at constant temperature, for any given mass of gaseous reactants there is a distribution of energies, with some particles having more energy than others.

A reaction can occur only if the particles have energy greater than, or equal to, the activation energy, $E_a$ (page 213).

The enthalpy-profile diagram in Figure 11.11 shows the activation energy for an exothermic reaction.

 Activation energy is the minimum amount of energy required for a reaction to occur.

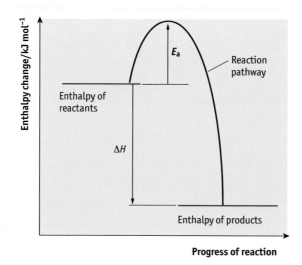

*Figure 11.11*
*Activation energy for an exothermic reaction*

Ludwig Boltzmann showed mathematically that the energies of molecules at a constant temperature are distributed as shown in Figure 11.12. This type of graph is called a **Boltzmann distribution**.

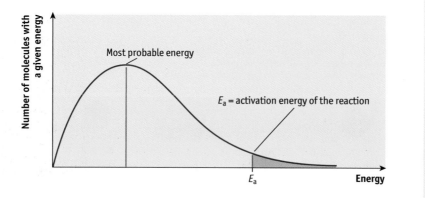

*Figure 11.12*
*Boltzmann distribution of energies*

- The distribution curve goes through the origin, which shows that there are no particles with zero energy.
- At high energy, the distribution approaches, but never touches, the horizontal axis. This shows that there is no maximum energy.
- The area under the curve represents the *total* number of particles (Figure 11.13).
- The shaded area represents the number of particles with energy greater than or equal to the activation energy, $E \geq E_a$. This shaded area, therefore, indicates the number of particles with sufficient energy to react.

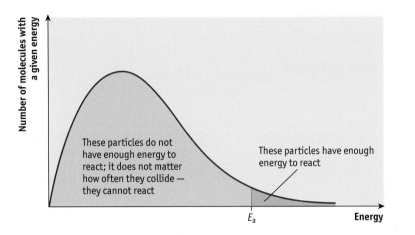

*Figure 11.13*

A rise in temperature increases the energy of the particles and, therefore, changes the number of particles with energy greater than or equal to the activation energy. This can be shown by comparing a typical distribution of energies at an initial temperature, $T_1$, with the distribution of energies when the temperature is raised to $T_2$ (Figure 11.14).

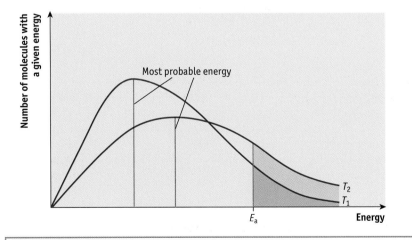

*Figure 11.14 Effect of temperature on the Boltzmann distribution ($T_2 > T_1$)*

### Ludwig Boltzmann, 1844–1906
Boltzmann made many contributions to the development of thermodynamics. Prone to bouts of depression, he committed suicide in 1906. His grave in Vienna is engraved with one of the most important thermodynamic equations that he derived: $S = k \log W$.

The number of particles remains the same. Therefore, the area under both curves is the same. However, at the higher temperature ($T_2$), the distribution flattens and shifts to the right, which means that there are fewer particles with low energy. The most probable energy is higher and a greater proportion of particles have energy that exceeds the activation energy.

Raising the temperature increases the number of particles with energy greater than or equal to the activation energy. This means that at higher temperatures there are more particles with sufficient energy to react. Therefore, there are more successful collisions, so the reaction is faster.

Lowering the temperature has the opposite effect.

## Effect of the use of a catalyst

Catalysts speed up reactions without themselves being changed permanently. They are divided into two types: **homogeneous** and **heterogeneous.** Both work by providing an alternative route, or mechanism, for the reaction that has a lower activation energy. This is illustrated in Figure 11.15 using an enthalpy profile and a Boltzmann distribution. On these diagrams, $E_a$ is the activation energy of the uncatalysed reaction; $E_{cat}$ is the activation energy of the catalysed reaction.

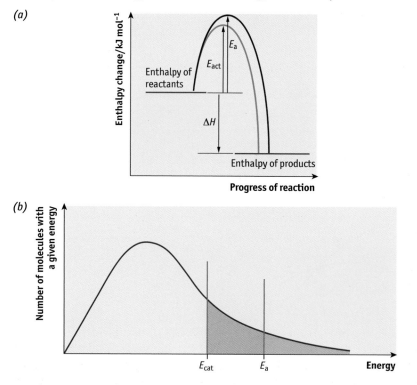

*(a)*

*(b)*

*Figure 11.15*
*(a) Enthalpy-profile diagram and*
*(b) Boltzmann distribution showing the effect of a catalyst on activation energy*

◀ The blue area indicates the proportions of particles with energy greater than the activation energy without the catalyst. The red and blue areas indicate the proportion of particles with energy greater than the activation energy with a catalyst.

A catalyst lowers the activation energy but does not alter either the enthalpy change for the reaction or the Boltzmann distribution. However, significantly, it increases the number of particles with energy greater than or equal to the new activation energy, $E_{cat}$. Therefore, the reaction rate is increased.

## Homogeneous catalysts

A **homogeneous catalyst** is in the same phase (gas, liquid or solid) as the reactants. This is often liquid because many reactions are carried out in aqueous solution. An example of a reaction that involves a homogeneous catalyst is the esterification (page 173) of ethanol and ethanoic acid (both liquids); concentrated sulphuric acid (a liquid) is the catalyst:

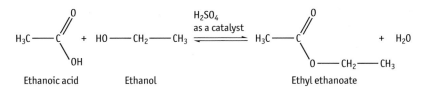

| Ethanoic acid | Ethanol | | Ethyl ethanoate |

The acid catalyst, $H^+$, provides an alternative mechanism that involves an intermediate step of lower activation energy. The $H^+$ ions from the catalyst take part in the reaction, but are released at the end of the reaction, so the final amount of the acid catalyst is the same as when the reagents were mixed.

Another example of homogeneous catalysis occurs in the loss of ozone from the upper atmosphere (stratosphere). This is a gas-phase reaction in which chlorine free radicals act catalytically (see page 231).

## Heterogeneous catalysts

A **heterogeneous catalyst** is in a different phase from the reactants. The most common type of heterogeneous catalysis involves reactions of gases in the presence of a solid catalyst.

The mode of action of a heterogeneous catalyst is different from that of a homogeneous catalyst. A heterogeneous catalyst works by adsorbing the gases onto its solid surface. This adsorption weakens the bonds within the reactant molecules, which lowers the activation energy for the reaction. Bonds are broken and new bonds are formed. The product molecules are then desorbed from the surface of the solid catalyst.

Transition metals are often used as heterogeneous catalysts. Iron is the catalyst in the Haber process for the manufacture of ammonia (page 222). The iron is either finely divided (and therefore has a large surface area) or in a porous form containing a small amount of metal oxide promoters:

$$N_2(g) + 3H_2 \overset{Fe(s) \text{ as catalyst}}{\rightleftharpoons} 2NH_3(g)$$

The production of poly(ethene) and other polymers from alkenes requires the use of a Ziegler–Natta catalyst. This is a mixture of titanium(IV) chloride and an organic compound of aluminium, $Al_2(CH_3)_6$. The mode of action of such catalysts is complicated and not altogether understood. However, it provides a surface for the adsorption of ethene.

Another important heterogeneous catalyst is the alloy of platinum, palladium and rhodium used in catalytic converters (page 233).

Catalysts are of great importance in the chemical industry.

1 Define the term 'activation energy' ($E_a$).

2 If the temperature is changed, the distribution of molecular energies changes.
   a Copy the distribution curve below and add two new curves:
      (i) at temperature $T_2$, which is higher than $T_1$
      (ii) at temperature $T_3$, which is lower than $T_1$

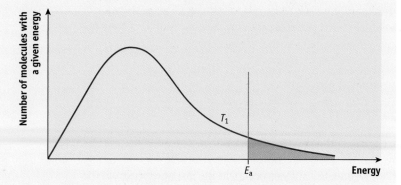

   b What do the three curves at $T_1$, $T_2$ and $T_3$ have in common?
   c Explain why increasing the temperature increases the rate of reaction.

3 a Explain what is meant by the term 'catalyst'.
   b Use a Boltzmann distribution curve to explain how a catalyst works.
   c Explain what is meant by homogeneous catalysis. Give an example.
   d Explain what is meant by heterogeneous catalysis. Give an example.

4 Describe how, and explain why, increasing the pressure on a gaseous reaction affects the rate of reaction.

# Chemical equilibrium

## Reversible reactions

There are many everyday examples of reactions or processes that are reversible, the best known being the changes of the physical states of water. When the temperature of water falls below 0°C, it freezes and ice forms. When the temperature rises above 0°C, the ice melts and water forms again. This process can be represented as:

$$H_2O(s) \rightleftharpoons H_2O(l)$$

The symbol $\rightleftharpoons$ is an equilibrium arrow and indicates that a reaction is reversible. Another reversible reaction is esterification (pages 173–74):

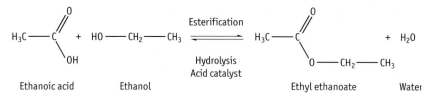

In the presence of an acid catalyst, ethanoic acid reacts with ethanol to produce the ester ethyl ethanoate and water. Ethyl ethanoate is hydrolysed by water, in the presence of an acid catalyst, to produce ethanoic acid and ethanol.

### Dynamic equilibrium

In the reversible reaction shown above, the reaction from left to right is called the **forward reaction**; the reaction from right to left is the **reverse reaction**.

If ethanoic acid and ethanol are refluxed in the presence of an acid catalyst, the forward reaction is initially fast because the concentrations of both reagents are high. However, as they react, the concentration of each reagent decreases, which lowers the rate of the forward reaction.

The reverse reaction is initially slow because the amount of ethyl ethanoate and water present is small. However, the concentrations of ethyl ethanoate and water build up, increasing the rate of the reverse reaction.

In summary, the forward reaction starts rapidly but slows down; the reverse reaction starts slowly and speeds up. It follows that a point is reached when the rate of the forward reaction equals the rate of the reverse reaction (Figure 11.16). When this happens, the system is in **dynamic equilibrium**.

- The system is in **equilibrium** because the amount of each chemical remains constant.
- The equilibrium is **dynamic** because the reactants and the products are both constantly interacting.

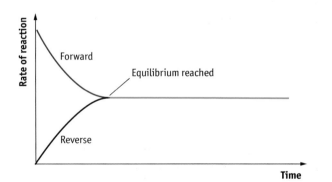

*Figure 11.16 Graph showing how rates of reaction change until equilibrium is established*

Dynamic equilibrium can only be achieved in a closed system (i.e. no reagent is allowed to escape).

A visual representation of a dynamic equilibrium is a 'Mexican wave', often seen at major sporting events. The wave moves around the stadium (dynamic), but as one person stands to wave, another sits down (equilibrium). The number of people waving and the number of people sitting is constant but the individuals are constantly changing.

> A dynamic equilibrium is reached when the rate of the forward reaction equals the rate of the reverse reaction. The concentrations of the reagents and products remain constant; the reactants and the products react continuously.

# Le Chatelier's principle

The French chemist Henri Le Chatelier studied dynamic equilibria and suggested a general qualitative rule that could be used to predict the movement of the position of the equilibrium.

> Le Chatelier's principle states that if a closed system at equilibrium is subject to a change, the system will move to minimise the effect of that change.

### Henri Le Chatelier, 1850–1836
Le Chatelier's principle led the way to the invention of the Haber process for the production of ammonia. In fact, Le Chatelier nearly invented it himself, but during his research he allowed air into the reaction chamber that contained the nitrogen and hydrogen and the apparatus exploded. This led him to abandon the project.

The factors that can be readily changed are concentration, temperature and pressure:

- If the concentration of a component is increased, the system moves to decrease the concentration of that component.
- If the temperature is increased, the system moves to decrease the temperature.
- If the pressure is decreased, the system moves to increase the pressure.

### Effect of changing concentration on the equilibrium position

Le Chatelier's principle applies to all reactions at equilibrium. However, there are few reactions in which the principle can be observed in action. An exception is the equilibrium between the yellow chromate ion, $CrO_4^{2-}$, and the orange dichromate ion, $Cr_2O_7^{2-}$ (Figure 11.17). Each ion is coloured, so it is possible to observe the movement of the position of equilibrium.

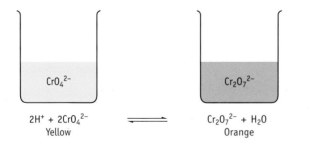

$$2H^+ + 2CrO_4^{2-} \rightleftharpoons Cr_2O_7^{2-} + H_2O$$

Yellow               Orange

*Figure 11.17*
*Equilibrium between the chromate ion, $CrO_4^{2-}$, and the dichromate ion, $Cr_2O_7^{2-}$*

Adding acid to the chromate(VI)/dichromate(VI) mixture increases the concentration of the hydrogen ions, $H^+(aq)$. The system now moves to minimise the effect, i.e. it tries to decrease the concentration of acid, $H^+(aq)$. This is achieved by the additional $H^+(aq)$ reacting with some of the $CrO_4^{2-}(aq)$ ions to form the products, $Cr_2O_7^{2-}(aq)$ and $H_2O(l)$. The position of the equilibrium moves to the right, so the mixture becomes more orange.

This can be reversed by adding a small amount of base, $OH^-(aq)$. The base reacts with the $H^+(aq)$, which decreases the concentration of the $H^+(aq)$ on the left-hand side of the equilibrium. The system moves to minimise the effect, i.e. it tries to increase the concentration of the acid, $H^+(aq)$. This is achieved by the $H_2O(l)$ reacting with some of the $Cr_2O_7^{2-}(aq)$ ions to form $CrO_4^{2-}(aq)$ ions and $H^+(aq)$ ions. The equilibrium position moves to the left and the orange solution becomes yellow.

### Large-scale application of Le Chatelier's principle

Ammonia is manufactured by the Haber process:

$$N_2(g) + 3H_2(g) \rightleftharpoons 2NH_3(g)$$

The boiling points of each of the gases are shown in Table 11.3.

| Gas | Boiling point |
|-----|---------------|
| Nitrogen | 77 K (−196°C) |
| Hydrogen | 20 K (−253°C) |
| Ammonia | 240 K (−33°C) |

*Table 11.3 Boiling points of nitrogen, hydrogen and ammonia*

If the equilibrium mixture is cooled to about −233 K (−40°C), the ammonia gas liquefies. Therefore, ammonia gas is lost from the equilibrium mixture. The system moves to minimise the effect of this loss. Nitrogen and hydrogen react to produce more ammonia gas to replace the ammonia gas that was liquefied and removed from the system.

### Effect of changing pressure on the equilibrium position

Changing pressure has almost no effect on the reactions of solids or liquids. It affects gaseous reactions only.

The pressure of a gas mixture depends on the number of gas molecules in the mixture. The greater the number of gas molecules, the greater is the pressure of the equilibrium mixture. If the pressure is increased in a system at equilibrium, the equilibrium position alters to decrease the pressure by reducing the number of gas molecules in the system. The equilibrium position is moved to the side that has fewer molecules.

If the pressure is increased on a system such as $2SO_2(g) + O_2(g) \rightleftharpoons 2SO_3(g)$, the equilibrium position moves to the right. This is because there are three molecules ($2SO_2 + 1O_2$) on the left-hand side of the equilibrium and two molecules ($2SO_3$) on the right-hand side. As a result of this shift in equilibrium position, the number of molecules is reduced and there is a reduction in pressure.

If the pressure is increased on a system such as $N_2O_4(g) \rightleftharpoons 2NO_2(g)$, the equilibrium position moves to the left, so that the number of molecules is reduced. This has the effect of reducing the pressure.

If the pressure is increased on a system such as $2HI(g) \rightleftharpoons H_2(g) + I_2(g)$, the position of the equilibrium does not move because there are equal numbers of gaseous molecules on each side of the equilibrium. A change in the equilibrium position does not affect the pressure.

### Effect of changing temperature on the equilibrium position

Temperature influences the rate of reaction and also plays an important role in determining the equilibrium position. The effect of temperature can be predicted only if the $\Delta H$ value of the reaction is known.

Consider a general equilibrium in which the forward reaction has $\Delta H$ of $-100\,kJ\,mol^{-1}$:

$$2A(g) + B(g) \rightleftharpoons C(g) + D(g) \quad \Delta H = -100\,kJ\,mol^{-1}$$

This means that the forward reaction is exothermic:

$$2A(g) + B(g) \rightarrow C(g) + D(g) \quad \Delta H = -100\,kJ\,mol^{-1}$$

The reverse reaction is endothermic:

$$C(g) + D(g) \rightarrow 2A(g) + B(g) \quad \Delta H = +100\,kJ\,mol^{-1}$$

If the temperature of the reaction mixture is increased, according to Le Chatelier's principle, the system alters to lower the temperature. This is achieved by the system favouring the reverse reaction, which is endothermic. This, therefore, helps to remove the additional enthalpy introduced by increasing the temperature. The position of the equilibrium moves to the left.

Decomposition of hydrogen iodide is an endothermic reaction:

$$2HI(g) \rightleftharpoons H_2(g) + I_2(g)$$

- If the temperature is increased, the equilibrium moves to the right.
- If the temperature is decreased, the equilibrium moves to the left.

Oxidation of sulphur dioxide is an exothermic reaction:

$$SO_2(g) + O_2(g) \rightleftharpoons 2SO_3(g)$$

- If the temperature is increased, the equilibrium moves to the left.
- If the temperature is decreased, the equilibrium moves to the right.

### Effect of using a catalyst on the equilibrium position

A catalyst is a substance that speeds up the rate of reaction by providing an alternative route, or mechanism, that has a lower activation energy. A catalyst does not alter the amount of product.

In a system at equilibrium, a catalyst speeds up the forward and the reverse reactions equally. Therefore, a catalyst has no effect on the *position* of the equilibrium.

However, catalysts play an important part in reversible reactions because they reduce the time taken to reach equilibrium. In the presence of a catalyst, the same amount of product is produced more quickly.

### The Haber process

Large amounts of nitrogenous compounds, particularly fertilisers, are needed by humans. Atmospheric nitrogen is in plentiful supply but it cannot be used directly. It has to be converted (fixed) into a usable compound. The Haber process 'fixes' nitrogen, converting it into ammonia:

$$N_2(g) + 3H_2(g) \rightleftharpoons 2NH_3(g) \quad \Delta H = -93\,kJ\,mol^{-1}$$

Le Chatelier's principle allows the optimum conditions for this industrial process to be determined.

The enthalpy change is −93 kJ mol⁻¹, so the forward reaction is exothermic. Therefore, if the temperature is increased, the equilibrium moves to the left and less ammonia is produced. It follows that a low temperature is optimum for the formation of ammonia.

*Figure 11.18 Production of ammonia by the Haber process*

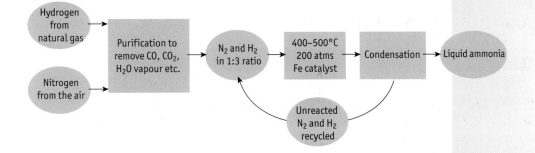

If the pressure is increased on the system $N_2(g) + 3H_2(g) \rightleftharpoons 2NH_3(g)$, the equilibrium position moves to the right to reduce the number of molecules. This has the effect of reducing the pressure. Therefore, high pressure is optimum for the formation of ammonia.

The effect of changing temperature and pressure on the yield of ammonia is shown in Table 11.4.

| Temp/K | 25 atm | 50 atm | 100 atm | 200 atm |
|--------|--------|--------|---------|---------|
| 373 | 92 | 94 | 96 | 98 |
| 573 | 28 | 40 | 53 | 67 |
| 773 | 3 | 6 | 11 | 18 |

*Table 11.4 Percentage yield of ammonia under different conditions*

The lowest temperature (373 K = 100°C) gives the highest percentage yield. However, at low temperature, the reaction rate is slow. A compromise has to be reached between yield and rate of reaction.

The highest pressure (200 atm) gives the highest percentage yield and also increases the reaction rate. However, at high pressure the operating costs increase. A compromise has to be reached between yield/rate and costs.

## Fritz Haber, 1868–1934

Ammonia is the basis of the manufacture of fertilisers and explosives. It has been suggested that, if the Haber process had not been invented, Germany would have been unable to maintain the supply of food and ammunition to troops during the First World War and that this could have meant an end to hostilities in 1915. Nevertheless, the process was of such importance that Haber was awarded a Nobel prize in 1918.

### Manufacturing conditions

The manufacture of ammonia in a modern plant is highly efficient. The operating conditions vary but are typically:

- a temperature of around 700 K (427°C)
- a pressure of around 100 atm

These conditions are a compromise.

The rate of reaction is increased by using a catalyst of either finely divided iron, or porous iron incorporating metal oxide promoters.

## Questions

**1 a** Explain what is meant by the term 'reversible reaction'.
   **b** Explain what is meant by the term 'dynamic equilibrium'.
   **c** State four external variables that could be changed to affect an equilibrium.

**2 a** State Le Chatelier's principle.
   **b** Use Le Chatelier's principle to deduce what happens to the following equilibrium when it is subjected to the changes stated:

$$2NO_2(g) \rightleftharpoons N_2O_4(g) \quad \Delta H = -57.2\,kJ\,mol^{-1}$$

   **(i)** the temperature is increased
   **(ii)** the pressure is decreased
   **(iii)** the $N_2O_4(g)$ is removed

**3** When the indicator methyl orange is dissolved in water, the following dynamic equilibrium is set up:

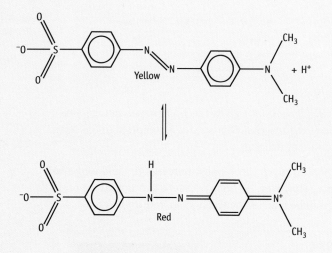

   **a** Hydrochloric acid is added to the equilibrium mixture. State, and explain, the colour change that occurs.
   **b** Then, aqueous potassium hydroxide, KOH(aq), is added dropwise to the solution in **a** until no further colour change occurs. Deduce the colour changes that occur. Explain your answer.

# Summary

You should now be able to:

- define enthalpies of reaction, formation, combustion and neutralisation
- recall what is meant by standard conditions
- draw an energy profile showing activation energy and use it to describe an exothermic or endothermic reaction
- describe simple laboratory experiments to determine enthalpy changes
- use the equation $Q = mc\Delta t$ to calculate enthalpy changes
- define Hess's law and understand its use in calculating unknown enthalpy changes
- define bond enthalpy
- use bond enthalpies to determine enthalpy changes of reaction
- understand the effect on the rate of a reaction of changing the concentration or pressure of reactants
- understand the effect of temperature on the rate of reaction and explain it using the Boltzmann distribution and an energy profile
- explain the role of a catalyst and give examples of homogeneous and heterogeneous catalysis
- understand what is meant by a reversible reaction and dynamic equilibrium
- state Le Chatelier's principle and use it to predict the effects of changes in temperature and pressure on a reversible reaction
- understand that in industrial processes there is a need to balance the effects of rate and equilibrium to establish optimum conditions

# Additional reading

Bond enthalpies can be used to decide whether reactions between covalent molecules are exothermic or endothermic. Exothermic reactions are more likely to occur, although it must be remembered that:

- if the reaction has a high activation energy, it may be difficult for the reaction to proceed
- there may be reasons connected with the way energy is distributed that make the reaction impossible

> ⓔ The second point is covered at A2.

Nevertheless, bond enthalpies provide some useful fundamental information.

Bond enthalpies are easily determined from enthalpies of formation. However, care must be taken to remember that these refer to the breaking of bonds in gases. For example, the enthalpy of formation of methane is $-75\,kJ\,mol^{-1}$ and is represented by the equation:

$$C(s) + 2H_2(g) \rightarrow CH_4(g)$$

Therefore, the enthalpy change of the reverse reaction to form $C(s)$ and $2H_2(g)$ from methane is $+75\,kJ\,mol^{-1}$. However, this figure does not represent the energy required to break four C–H bonds because hydrogen and solid carbon (graphite) are formed. To obtain the C–H bond energy, an enthalpy cycle is required that includes the enthalpy change when 1 mol of $H_2(g)$ bonds is broken and the enthalpy change when 1 mol of $C(s)$ is converted to 1 mol of $C(g)$ (the enthalpy of atomisation):

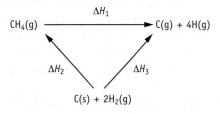

$\Delta H_1 = 4 \times$ (bond energy of the C–H bond)
$\Delta H_2 = $ enthalpy of formation of $CH_4(g) = -75\,kJ\,mol^{-1}$

$\Delta H_3 = 2 \times$ (bond energy of $H_2$) + enthalpy of atomisation of $C(s)$
$\qquad = (2 \times 436) + 715 = 1587\,kJ\,mol^{-1}$
$\Delta H_1 = \Delta H_3 - \Delta H_2$
$\qquad = 1587 - (-75) = 1662\,kJ\,mol^{-1}$

Therefore, the bond enthalpy of a C–H bond is $\dfrac{1662}{4}$ $= 415.5\,kJ\,mol^{-1}$. This value can be used to calculate other bond enthalpies, at least approximately. (Remember that the figure obtained is an *average* figure that does not take into account the effect of other surrounding bonds.) This is illustrated by the following cycle to calculate the strength of a C–C bond, based on ethane:

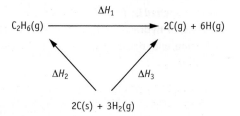

For this cycle:

$\Delta H_1 = (3 \times 436) + (2 \times 715) - (-84.6) = 2822.6\,kJ\,mol^{-1}$

This figure represents the bond enthalpies of six C–H bonds plus one C–C bond.

enthalpy of the C–C bond $= 2822.6 - (6 \times 415.5)$
$\qquad\qquad\qquad\qquad = 329.6\,kJ\,mol^{-1}$

This is not the generally accepted figure ($347\,kJ\,mol^{-1}$), but it is close enough.

The same process can be used to build up information on other bonds.

Bond enthalpies can be used to explain why carbon is uniquely able to form stable chains and rings with itself, whereas silicon (also in group 4) can do so only by incorporating oxygen:

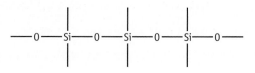

Rings containing Si–O are common. Soils and rocks contain these units, built into complex three-dimensional structures.

The relevant bond enthalpies for carbon and silicon are shown in Table 11.5.

**Table 11.5**

| Bond | Enthalpy/ kJ mol$^{-1}$ | Bond | Enthalpy/ kJ mol$^{-1}$ |
|------|------|------|------|
| C–C | 347 | Si–Si | 226 |
| C–H | 413 | Si–H | 318 |
| C–O | 358 | Si–O | 466 |

It can be seen that the Si–Si bond is a weak bond, whereas the Si–O bond is much stronger. The C–C bond is stronger than the Si–Si bond and is similar in strength to a C–O bond.

The differing stabilities of carbon and silicon compounds can be related to their electron arrangements. Elements in the second period do not have empty orbitals low enough in energy to react with a nucleophile. This is not the case for elements in the third period, which allows greater opportunities for nucleophilic attack to take place. This means that some compounds of silicon are readily converted to more stable products.

## Extension question: Apollo 11

The *Apollo 11* lunar module was the first manned spacecraft to land on the moon, in July 1969. It was propelled by Aerozine 50, a 50:50 mixture, by mass, of hydrazine ($N_2H_4$) and 1,1-dimethyl-hydrazine, with dinitrogen tetroxide ($N_2O_4$) as an oxidising agent.

Apollo 11 leaves the launch tower

NASA/SPL

The products of the reaction between Aerozine 50 and dinitrogen tetroxide are nitrogen, water and carbon dioxide.

a  Write an equation for the reaction between hydrazine and dinitrogen tetroxide.

b  Write an equation for the reaction between 1,1-dimethylhydrazine and dinitrogen tetroxide.

c  If 2200 kg of Aerozine 50 is required for the lunar module to take off, what mass of oxidising agent is needed?

d  Assuming complete oxidation of the fuel, use the following standard enthalpies of formation to calculate the enthalpy released at take-off.

$$\Delta H^{\ominus}_f(N_2H_4) = 50.6\,\text{kJ mol}^{-1}$$
$$\Delta H^{\ominus}_f((CH_3)_2NNH_2) = 49.3\,\text{kJ mol}^{-1}$$
$$\Delta H^{\ominus}_f(H_2O(g)) = -241.8\,\text{kJ mol}^{-1}$$
$$\Delta H^{\ominus}_f(N_2O_4) = -19.5\,\text{kJ mol}^{-1}$$
$$\Delta H^{\ominus}_f(CO_2) = -393.7\,\text{kJ mol}^{-1}$$

e  A swimming pool contains 350 000 dm$^3$ of water. Suppose the enthalpy generated at take-off was used to heat this water, which had an initial temperature of 20 °C. What would be the final temperature reached? Ignore heat losses to the surroundings and assume that the specific heat capacity of the water is 4.2 J g$^{-1}$ K$^{-1}$ and that its density is 1 g cm$^{-3}$.

# Resources

## Chapter 12

There is a growing consensus that use of the Earth's natural resources should be controlled. Better ways of providing energy are being sought, not just to conserve the limited supplies of oil and coal but also to reduce the pollution caused by burning them. There is concern over the quality of the air that we breathe and over the longer-term effects that changes to the composition of the atmosphere may have on the climate.

This chapter examines the greenhouse effect and the role that radicals have in damaging the ozone layer. Chemists have an important role in explaining these phenomena and in suggesting ways to reduce the likely adverse effects. The steps that are being taken to control polluting gases are also addressed. The movement towards establishing more efficient industrial processes and the principles of 'green chemistry' are reviewed.

# Infrared absorption and the greenhouse effect

The Earth is warmed mostly by energy transmitted from the Sun. This energy consists largely of visible light, but there is also some ultraviolet and infrared radiation. Most ultraviolet radiation is removed in the upper parts of the atmosphere (the stratosphere) by the ozone layer. Of the remaining radiation, some is reflected and some is absorbed by the atmosphere. Carbon dioxide and water molecules remove part of the incoming infrared radiation. About 50% of the energy from the Sun arrives at the Earth's surface.

Once at the Earth's surface, chemical reactions absorb and transmit this energy. Over time, the surface temperature of the Earth has remained approximately constant because an equilibrium has been established between arriving and departing energy. The departing energy — which is almost wholly infrared radiation — does not pass unhindered into outer space. If it did, the Earth would be a very cold place. Many gases in the atmosphere absorb some of this energy and reflect it back to Earth. In other words, these gases act like a thermal blanket. Carbon dioxide and water molecules are not the only molecules to behave in this way. Gases that have bonds that vibrate in a way that alters the electrical balance of the molecule also absorb infrared radiation.

## Greenhouse gases

Of the gases naturally present in the atmosphere, neither nitrogen nor oxygen absorb infrared radiation. Gases that do absorb infrared are collectively known as **greenhouse gases**. There is a wide range of greenhouse gases, since most molecules contain polar bonds.

### Contribution of a gas to the greenhouse effect

The overall contribution of a gas to the greenhouse effect depends on:

- its ability to absorb infrared radiation
- its atmospheric concentration

These factors can be treated quantitatively. The relative greenhouse effect of four gases is shown in Table 12.1.

| Gas | Formula | Approximate relative greenhouse effect | Approximate atmospheric concentration (%) |
|---|---|---|---|
| Carbon dioxide | $CO_2$ | 1 | 0.035 |
| Methane | $CH_4$ | 25 | 0.00017 |
| Dinitrogen oxide | $N_2O$ | 250 | 0.00003 |
| CFC-12 | $CCl_2F_2$ | 25 000 | $\sim 4 \times 10^{-8}$ |

*Table 12.1*
*Contribution of gases to the greenhouse effect*

Table 12.1 shows that there are large differences between the contributions to the greenhouse effect made by individual molecules. However, this must be set against the amount of each gas present in the atmosphere. Carbon dioxide has the highest concentration, whereas that of CFC-12 is low and, as CFCs are phased out, will become lower. A major source of atmospheric methane is the reduction of carbon-containing compounds under anaerobic conditions. It may cause amusement that this occurs in the digestive tract of cows, but it is also a hazard of the decomposition of rubbish in compacted landfill sites. Dinitrogen monoxide is released as a result of the reduction of nitrates on agricultural land.

Each of the bonds in the greenhouse-gas molecules (C=O in carbon dioxide, C–H in methane, O–H in water and so on) absorbs infrared radiation of a particular wavelength and vibrates with increased energy. This energy is then randomly dispersed, with much of it returning to the Earth's surface.

### Effect of greenhouse gases

As the concentration of greenhouse gases rises, the average temperature at the Earth's surface will increase. It is this increase that is usually referred to as the greenhouse effect. However, it is more accurate to call it the 'enhanced greenhouse effect' because it is the *extra* heating that is the cause for concern.

The role of carbon dioxide as a contributor to **global warming** is the focus of much attention. This is not because it is the most effective greenhouse gas nor that it is a major contributor, but because it is produced in huge quantities by burning fuels such as wood, coal and oil products. (In fact, water vapour is the biggest contributor to the greenhouse effect.) Many attempts have been made to predict the result of continued emissions of carbon dioxide into the atmosphere at the current rate. There is no universal agreement, but there is serious concern that it could result in the melting of the polar ice-caps, causing extensive flooding of low-lying land, and that changing temperature patterns could lead to severe droughts in some parts of the world.

*e* Remember that charge separation due to the differing electronegativities of atoms is an important feature of most covalent bonds.

*e* There is a change in oxidation number from +5 in nitrates to +1 in dinitrogen monoxide.

The difficulty of providing a reliable prediction emphasises the complex nature of the chemistry of the atmosphere. Animals release carbon dioxide when they respire; plants absorb carbon dioxide during photosynthesis. Much carbon dioxide is absorbed as it dissolves into surface waters. We depend on forests to reduce the level of carbon dioxide and it is not clear how trees would be affected if the global temperature continues to rise. In many countries, extensive deforestation is also taking place.

In addition, it is not possible to predict how long an individual molecule may stay in the atmosphere — only average figures are available (residence times). These factors make it difficult to predict accurately the effect of increasing pollution.

Scientists have an essential role to play. They should:
- try to extend the knowledge of the processes involved, to allow predictions to be made with the greatest certainty possible
- make their findings widely known, so that governments can understand the potential impact
- use initiatives, such as the Kyoto protocol, to monitor progress towards a cleaner environment
- research alternative sources of power and processes to manufacture chemicals that have less environmental impact
- search for solutions to resolve current pollution issues

Not all the points in the above list might be considered as part of the traditional role of scientists, but it is becoming increasingly clear that communication at all levels is essential to solve problems that extend beyond national boundaries.

## Carbon dioxide removal

It is not always easy to avoid the release of unwanted gases into the atmosphere. However, in the case of carbon dioxide, the technology exists to make a significant contribution towards its control.

Carbon dioxide does not readily exist as a liquid. Solid carbon dioxide normally sublimes (i.e. changes from solid to gas without passing through the liquid stage).

**e** This property makes it useful for creating misty vapours in theatrical productions.

Nevertheless, carbon dioxide can be liquefied by changing the conditions of temperature and pressure. The necessary conditions are called **supercritical**. A scheme that is gaining widespread approval is **carbon capture and storage** (CCS). Gaseous carbon dioxide is collected and compressed. Then, using an appropriate temperature and pressure, it is converted into a liquid. The liquid carbon dioxide can be injected deep into the oceans for storage. Possible sites include depleted oil and gas fields and unmineable coal seams. The method might only be appropriate for large-scale emitters of carbon dioxide such as power stations and industrial plants, but these do contribute significantly to the overall release of the gas. The risk is the possible subsequent leakage of the gas, but in trials this has not proved to be a problem. This method has been used in Norway for a number of years.

An alternative approach could be to react carbon dioxide with metal oxides to form carbonates. However, there is currently no practical way of doing this on a large scale.

Both methods are expensive, but the UK government is making money available to tackle the problem.

## The ozone layer

Oxygen normally exists as an $O_2$ molecule, but oxygen atoms can also combine to form ozone, $O_3$. The structure of ozone can be represented as follows:

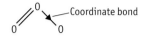

Ozone is a reactive gas that does not occur normally in the atmosphere close to the Earth's surface. However, some 25 km higher up, in the stratosphere, there is a layer of ozone approximately 15 km deep. This may sound substantial, but it would have a depth of no more than 3 mm if it existed at lower altitudes, because of the difference in atmospheric pressure. The ozone layer is important because it absorbs much of the ultraviolet radiation emitted by the Sun. If this were to reach the surface of the Earth, it would cause an increase in sunburn and skin cancers in humans.

In the stratosphere, ozone is in equilibrium with oxygen:

$$O_3 \rightleftharpoons O_2 + O$$

The oxygen radical, O, does not exist naturally in the laboratory, but occurs under the different conditions at high altitudes.

The forward reaction that breaks $O_3$ into $O_2$ and O is triggered by ultraviolet radiation of wavelength approximately 300 nm. The reverse reaction that results in the formation of ozone absorbs more energetic radiation of wavelength less than 242 nm.

This high energy is needed because the oxygen radicals have to be formed from oxygen molecules:

$$O_2 \rightarrow O + O$$

An oxygen radical then reacts with an oxygen molecule to produce a molecule of ozone:

$$O_2 + O \rightarrow O_3$$

### Causes of the hole in the ozone layer

It was a shock when it was discovered, during a polar spring season, that the ozone layer above the Antarctic had thinned to the point where a hole had developed. This discovery showed how important the management of the atmosphere is.

### Chlorine radicals

The main cause of the hole in the ozone layer is almost certainly chlorine radicals, created by the homolytic breakage of a covalent bond. They attack the ozone layer, as shown below.

### Mechanism of action of chlorine radicals

First, there is an initiation step resulting in the formation of a chlorine radical, Cl•. This is most likely to be the result of the breaking of a C–Cl bond.

$$R{-}Cl \rightarrow R\bullet + Cl\bullet$$

**e** You do not need to remember the wavelengths, but you should appreciate that a wide spectrum of ultraviolet light is absorbed in maintaining the equilibrium.

**e** You should appreciate the relationship between this mechanism and the reactions of alkanes with free radicals.

The chlorine radical is then involved in the propagation steps:

$$Cl\bullet + O_3 \rightarrow ClO\bullet + O_2$$
$$ClO\bullet + O \rightarrow Cl\bullet + O_2$$

In the second propagation step, the chlorine radical is regenerated and can cause the destruction of another ozone molecule, which will generate the production of another chlorine radical and so on. It has been estimated that approximately 1000 ozone molecules could be destroyed as the result of the production of a single chlorine radical.

The sequence is terminated only if the chlorine radical combines with another radical to remove it from the propagation sequence. For example, a chlorine molecule might be formed or a molecule of R–Cl recreated.

*Accumulation and release of chlorine-containing compounds*
The hole over the Antarctic was first observed in the spring because of the conditions in the polar regions during this season. During the winter months it is dark; clouds form and concentrate the chlorine-containing compounds. In the spring, sunlight reaches the clouds, which rapidly disperse, releasing the damaging chemicals in a destructive burst. Fortunately, because of the nature of equilibria, the ozone is regenerated and the layer is reformed eventually, but this takes time.

*Sources of chlorine-containing compounds*
One source of chlorine-containing chemicals in the atmosphere is CFCs (page 186). It is possible that other chlorine compounds are involved, as many have been used as solvents, for example in dry cleaning. However, these latter compounds are much more likely to have been destroyed in chemical reactions lower in the atmosphere, whereas CFCs have low chemical reactivity and are more stable.

The use of CFCs should by now have been phased out. However, alternatives such as tetrafluoroethane are also greenhouse gases.

The need to be aware of the possible side effects of the use of chemicals is becoming understood increasingly, by both the chemical industry and consumers.

### Nitrogen monoxide

Compounds that contain a C–Cl bond are not alone in supplying radicals that initiate reactions responsible for the destruction of the ozone layer. Of growing concern is the role of nitrogen monoxide. Thunderstorms are a major source of this gas, as lightning causes nitrogen and oxygen to combine:

$$N_2(g) + O_2(g) \rightarrow 2NO(g)$$

Nitrogen monoxide is also produced by the high-temperature combustion of nitrogen and oxygen that occurs to a small extent in internal combustion engines. More significantly, it is present in the exhaust gases of aircraft.

Complex reactions in the atmosphere may result in most of this gas being removed. For example, some may be oxidised and combine with moisture to create nitric acid, which is a contributor to acid rain. However, a small amount of nitrogen monoxide migrates to the stratosphere, where it attacks the ozone layer in a series of reactions analogous to those of the chlorine radical. There is no initiation step, because nitrogen monoxide already possesses a 'free' electron. The propagation steps are as follows:

$$NO + O_3 \rightarrow NO_2 + O_2$$
$$NO_2 + O \rightarrow NO + O_2$$

A comparison with the propagation steps for the chlorine free radical shows that these steps can be expressed more generally using 'R' to represent the radical:

$$R + O_3 \rightarrow RO + O_2$$
$$RO + O \rightarrow R + O_2$$

# Combustion and pollution

Global warming and the damage to the ozone layer are complex, long-term problems. Other forms of atmospheric pollution are more direct and immediate. Some gases emitted as by-products of industrial processes, the treatment of waste and the burning of fuels are toxic and may take part in complex reactions that result in the build-up of other undesirable pollutants.

## Internal combustion engines

The internal combustion engine is the source of many of the pollutants discharged into the atmosphere. However, catalytic converters have significantly reduced the level of the more immediately dangerous pollutants. Without a catalytic converter, a car would discharge several toxic gases into the atmosphere, some of which (e.g. benzene) are also carcinogenic.

Carbon monoxide is formed by the incomplete combustion of fuel:

$$C_8H_{18}(g) + 8\tfrac{1}{2}O_2(g) \rightarrow 8CO(g) + 9H_2O(g)$$

The internal combustion engine reaches temperatures of approximately 1000°C. This high temperature provides sufficient energy for nitrogen and oxygen (both present in air) to react to form nitrogen oxides. High in the atmosphere, nitrogen monoxide can survive longer than it does at the Earth's surface, where further oxidation to nitrogen dioxide occurs:

$$N_2(g) + O_2(g) \rightarrow 2NO(g)$$
$$2NO(g) + O_2(g) \rightarrow 2NO_2(g)$$

### Catalytic converters

Catalytic converters, fitted to all modern cars, reduce the emission of unburnt hydrocarbons, carbon monoxide and oxides of nitrogen by promoting the following reactions:

- removal of unburnt hydrocarbons:
$$C_8H_{18}(g) + 12\tfrac{1}{2}O_2(g) \rightarrow 8CO_2(g) + 9H_2O(g)$$
- removal of carbon monoxide and oxides of nitrogen:
$$2NO(g) + 2CO(g) \rightarrow N_2(g) + 2CO_2(g)$$
$$2NO_2(g) + 4CO(g) \rightarrow N_2(g) + 4CO_2(g)$$

A catalytic converter consists of a fine aluminium mesh coated with a thin solid film of an alloy of platinum, rhodium and palladium. Catalytic converters function efficiently only at high temperatures and they are, therefore, not particularly effective on short journeys, when the engine may not have warmed up sufficiently.

# Other environmental factors

As well as the gaseous products already covered, vehicles that run on diesel may produce fine particles of carbon in the exhaust fumes, which result from the complete breakdown of the fuel. These particles can cause problems for asthmatics.

Unburnt fuel released into the atmosphere, together with carbon monoxide, nitrogen monoxide, nitrogen dioxide and water vapour, forms a mixture that can undergo radical reactions. These reactions generate ozone, which is toxic, even at low concentrations. A further series of reactions then takes place, creating unexpected organic molecules, such as $CH_3COO_2NO_2$. This product is known by the old name peroxyacetylnitrate or PAN. It is an unpleasant compound that irritates mucous membranes and can cause severe breathing difficulties. It is formed only in the upper atmosphere, but occasionally the conditions are such that mixing of the atmosphere occurs. Then, this compound, along with nitrogen dioxide and ozone, occurs at sufficiently low altitude to become a severe risk. The conditions required are bright sunlight, still air and a high concentration of exhaust pollutants — for example, during morning rush hour in a big city. The unpleasant cocktail of gases is referred to as **photochemical smog**. It is characterised by the brown haze caused by the presence of high concentrations of nitrogen dioxide molecules.

◀ Nitrogen dioxide is a brown gas.

*Photochemical smog over Los Angeles*

It is now commonplace for air quality to be closely monitored and the results are readily accessible. The method of analysis employed is infrared spectroscopy (page 187). Infrared analysis is also used to monitor carbon monoxide emissions from cars.

# Green chemistry

Growing public awareness that the future well-being of our planet may depend on our ability to control environmental pollution has led to the development of **green chemistry**. There are principles that responsible scientists should follow that respect the limited nature of our resources and recognise the importance of addressing the effect that current technology will have on future generations:

- Industrial processes should minimise the production and use of hazardous chemicals.
- Processes should use renewable materials.
- Energy requirements should be minimised by the use of selective catalysts and sustainable energy sources.
- Products should be biodegradable.
- Processes should achieve a high atom economy (page 138) to avoid the unnecessary waste of materials.

Most of these principles may seem to be commonsense, but they have not been adhered to in the past. They present a formidable challenge to industry. Nevertheless, it is worth considering some examples where a greater understanding and more sophisticated technology have brought about progress towards a cleaner, safer environment.

The following examples illustrate how science confronts problems and how more environmentally appropriate solutions are being sought.

## Unleaded fuels

Tetraethyllead, $Pb(C_2H_5)_4$, used to be added to petrol to provide smoother combustion in car engines. Its function was to limit overheating, by conducting heat away from the point of combustion. The discovery that the lead released through the exhaust was toxic led scientists to modify the composition of the petrol. For example, a greater proportion of branched hydrocarbons was included, making the addition of tetraethyllead unnecessary. These improved fuels were more expensive to produce, but the extra cost to the consumer was offset by a reduction in fuel tax. It is a good example of progress made by a combination of scientific effort and political will.

### Thomas Midgley, 1889–1944

Thomas Midgley was an American, born in 1889. A prolific inventor, he had a talent for solving practical problems. Two of his discoveries were the use of lead tetraethyl in petrol and CFCs in refrigerators. Both were hailed as important breakthroughs at the time. The later recognition of their undesirable side effects illustrates the problems of the introduction of new materials. Midgley contracted polio at the age of 51 and designed an arrangement of ropes and pulleys to lift himself out of bed. By a cruel turn of fate, aged 55, he became entangled in his own rope system and strangled himself.

SPL

Lead has been largely eliminated from paint and electrical components. Lead piping is still commonly found in older houses but is being replaced by copper or plastic.

Fossil fuels are non-renewable. Replacement by renewable fuels such as ethanol and biodiesel is being actively researched (pages 150–52).

## Dry cleaning

Dry cleaning involves an organic liquid as the solvent, rather than water. Tetrachloromethane ($CCl_4$) was abandoned when it was found that it causes liver and kidney damage. It has been replaced by other chlorinated compounds, such as tetrachloroethene ($Cl_2C=CCl_2$). However, the use of any chlorinated chemical is undesirable, since its disposal is problematic and escape into the atmosphere is inevitable.

An alternative solvent is 'supercritical' liquid carbon dioxide, which is formed at a temperature of about 40°C and a pressure between 75 and 200 times normal atmospheric conditions. The cleaning process using this liquid is efficient and the excess carbon dioxide is readily absorbed by an alkali.

◀ Supercritical carbon dioxide is used in the foaming of polymers and as a solvent to extract caffeine from coffee beans and tea leaves.

## Fire extinguishers

Some fires can be extinguished using water, but those involving organic liquids are harder to deal with. Pouring water onto oil that has caught fire in a frying pan is both ineffective and dangerous because the oil floats on top of the water and continues to burn. This is because organic liquids and water do not mix, as there are no intermolecular forces to bind the molecules, and oils are less dense than water. Small-scale fires might be put out by smothering with a blanket of carbon dioxide and starving them of oxygen, but larger fires require a different approach. Halogenated organic liquids (e.g. CFCs) were originally used because they are non-flammable. Two examples are Halon-1211, $CBrClF_2$, and Halon-1301, $CBrF_3$.

◀ Under the Montreal Protocol the use of Halon-1211 and Halon-1301 has been banned since 1996.

One alternative to halogenated compounds is known as PYROCOOL®FEF. This mixture is a foaming agent; the foam smothers the fire and puts it out. PYROCOOL®FEF has other advantages:
- The bubbles that it forms collapse quickly and coat the surface.
- It is non-toxic and biodegradable.
- It contains chemicals that absorb the high energy from the fire and re-emit it at longer, less dangerous wavelengths, which helps stop the fire spreading.

## Global pollution: international action

Global pollution affects everyone and there has to be a broad political will to find solutions. There are several international agreements on pollution, but their conditions have not been accepted fully by every nation. Nonetheless, the Montreal Protocol on Substances that Deplete the Ozone Layer, agreed and signed by the EEC and other countries in 1987, was an important step towards the phasing out of CFCs, halons and other chlorine-containing chemicals. The Kyoto Protocol has led to targets for the reduction of greenhouse gases which, although proving difficult to meet, is a step in the right direction.

In 2001, a convention aimed at reducing or eliminating the use of certain organic chemicals was agreed by a number of countries. The properties of these substances, known as persistent organic pollutants (POPs), are that:
- they are not easily degraded
- they travel distances through the environment
- they are incorporated into the food chain

POPs include:
- some insecticides, such as DDT
- some compounds unintentionally produced during combustion, such as dioxins
- some components of paints and plastics

In 1982, in Rio de Janeiro, a more general agreement was made by the United Nations that affirms the importance of cooperation between nations to address 'environmental human rights'. It is a serious issue that countries producing air-borne pollution may not be the countries that suffer from its effects. The Rio Declaration was intended to establish principles for a basic framework to resolve some of the potential conflicts.

It is increasingly the case that scientists must accept that much of their work is not just concerned with the solution to problems but may carry additional responsibilities to the society in which they work, and to the wider community.

# Summary

You should now be able to:
- explain the greenhouse effect in terms of the effect of infrared radiation on covalent bonds
- explain that the contribution a gas makes to the greenhouse effect is affected by:
  - its ability to absorb infrared radiation
  - its concentration
- describe carbon capture and storage
- describe the chemical structure and role of the ozone layer
- use equations to explain how the ozone layer is destroyed by $Cl\bullet$ and NO radicals
- explain the release of unburned hydrocarbons, carbon monoxide and nitrogen oxides from the internal combustion engine
- explain environmental concerns over the release of unburned hydrocarbons, carbon monoxide and nitrogen oxides
- use equations to outline the action of a catalytic converter
- explain the principles that should govern good practice in the manufacture and use of chemicals
- outline the role of chemists in limiting the adverse effects of chemical products

# Additional reading

## Greenhouse gases

The impact of climate change on the global environment is relevant to us all and a key question is how human activity affects the environment. In an attempt to control the extent of atmospheric pollution, a number of inter-governmental treaties have been signed, including the Kyoto agreement. This aims to limit worldwide emissions of greenhouse gases, such as carbon dioxide, methane, dinitrogen oxide and ozone. Careful monitoring is essential if we are to understand fully the impact of elevated levels of greenhouse gases on the environment.

It is known that levels of carbon dioxide in the atmosphere have increased greatly in the last 50 years. Measurements of levels of carbon dioxide trapped in ice indicate that the pre-industrial carbon dioxide level was 278 parts per million (ppm). It varied by no more than 7 ppm during the 800 years between 1000 and 1800. Between 1958 and the end of 2004, the carbon dioxide level in the atmosphere increased from about 315 ppm to 378 ppm. As the pre-industrial level was 278 ppm, this means that human activities have increased the concentration of atmospheric carbon dioxide by 100 ppm, which is 36%. Concentrations of gases in the atmosphere are monitored by infrared spectroscopy.

## Infrared spectroscopy

Even in a solid, the atoms in molecules are not stationary or held in a fixed position. They vibrate so that the bonds between them stretch and compress and the molecules bend. Each vibration occurs at a specific frequency, which depends on the atoms in the bond and on the other atoms that make up the molecule.

All molecules absorb energy. When a molecule absorbs energy that corresponds exactly to the vibrational energy of a particular bond, the size (or amplitude) of vibration increases and the bond resonates in an 'excited state'. When the molecule returns to the ground state, the absorbed energy is emitted and can be detected. This energy is characteristic of the bond and the frequency lies in the infrared region of the electromagnetic spectrum.

The infrared absorption can be used to identify the bond only if the absorbed energy brings about a change in dipole moment. The dipole moment for a pair of opposite charges is defined as the magnitude of the charge multiplied by the distance between them. Many bonds possess a dipole moment because of the difference in electronegativity of the atoms creating the bond:

$$\frac{\text{dipole}}{\text{movement}} = \frac{\text{magnitude}}{\text{of the charge}} \times \frac{\text{distance between the charges}}{\text{(bond length)}}$$

In symmetrical molecules, there may be no overall dipole moment. This could be because there is no charge separation (as is the case in the diatomic molecules of elements) or because the dipole moments of individual bonds are cancelled out by the dipole moments of other bonds. When this happens, infrared is not absorbed. However, when stretching and/or bending occurs, the dipole moment may change. Polar bonds, such as C–O, C=O and O–H, give rise to large absorptions; non-polar bonds, such as C–C and C=C, produce only weak absorptions. Bending vibrations require less energy than stretching vibrations and, therefore, occur at lower energy, or lower wavenumber.

Planar molecules, such as sulphur dioxide and carbon dioxide, only absorb infrared energy by bond stretching or bond bending. In sulphur dioxide there are three possible vibrations:

- a symmetrical stretch — both bonds stretch at the same time

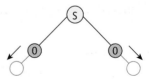

- an asymmetrical stretch — one bond stretches as the other compresses

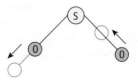

■ bending — the O–S–O bond angle changes

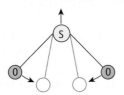

Each motion results in a change in dipole moment, so each absorbs infrared radiation. Therefore, the infrared spectrum of sulphur dioxide shows three different absorptions, as shown in Figure 12.1.

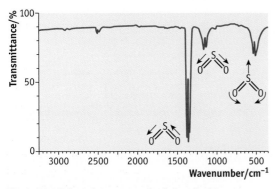

**Figure 12.1** *Infrared spectrum of sulphur dioxide*

Carbon dioxide also has three possible vibrations:
■ a symmetrical stretch — both bonds stretch at the same time

■ an asymmetrical stretch — one bond stretches as the other compresses

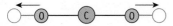

■ bending — the O–C–O bond angle changes

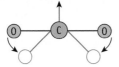

The carbon dioxide molecule is linear, so symmetrical stretching does not result in a change in dipole moment, because the effect on one bond is balanced by the effect on the other bond. Consequently the infrared spectrum of carbon dioxide shows only the vibrations due to asymmetrical stretching and bond bending, as shown in Figure 12.2.

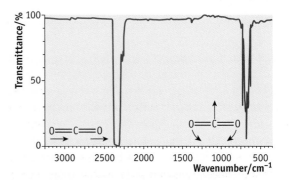

**Figure 12.2** *Infrared spectrum of carbon dioxide*

Larger molecules have a greater number of vibrational modes. It is possible to imagine different parts of molecules stretching, bending, swaying, twisting and rocking. Each vibration results in a distinctive absorption of infrared radiation. Some of the possible vibrational modes are shown in Figure 12.3. The three atoms shown are initially all in the same plane.

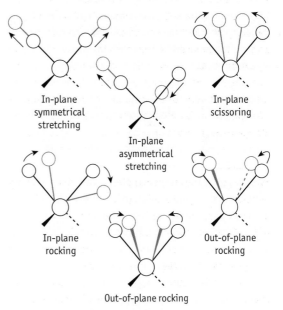

**Figure 12.3** *Possible vibrational modes*

With larger molecules, the number and variety of vibrations increase. Therefore, their infrared spectra have a large number of peaks. However, certain bonds always absorb in the same region, irrespective of the rest of the molecule. This means that infrared spectroscopy can be used to identify specific functional groups.

## Extension question: London to Edinburgh

There is ever-growing concern that not enough attention is being paid to atmospheric pollution. Gases such as carbon dioxide, sulphur dioxide, nitrogen oxides and even water vapour contribute to the greenhouse effect. These gases, and others, prevent the escape of heat from the Earth's surface and are thus contributing to a small, but potentially significant, rise in the temperature of the Earth's atmosphere.

There are a number of oxides of nitrogen. Nitrogen monoxide is of particular concern because, if it travels to the outer atmosphere (the stratosphere) it destroys ozone molecules. The Earth is protected by a layer of ozone in the stratosphere that absorbs harmful ultraviolet radiation that would otherwise be damaging to life. For example, skin cancers would become more common.

According to a recent news item, a medium-sized car travelling from London to Edinburgh (a distance of 400 miles) emits about 100 kg of carbon dioxide.

Assume that petrol — a mixture of hydrocarbons — can be represented as octane, $C_8H_{18}$.

a  Write an equation for the complete combustion of $C_8H_{18}$.

b  How many moles are there in 100 kg of carbon dioxide?

c  How many moles of $C_8H_{18}$ are used on the journey from London to Edinburgh?

d  What mass of $C_8H_{18}$ is burned on the journey from London to Edinburgh?

e  Calculate the fuel consumption of the car in miles per gallon. (The density of petrol is $0.737\,g\,cm^{-3}$ and 1 gallon = 4.546 litres ($dm^3$).)

The same journey taken in an aeroplane might result in the emission of 7500 kg of carbon dioxide (equivalent perhaps to 75 kg per passenger). The fuel used is kerosine, which, although it is a mixture, can be considered to be $C_{13}H_{28}$. The density of kerosine is $0.817\,g\,cm^{-3}$

f  Repeat the calculations, using $C_{13}H_{28}$, to estimate the average fuel consumption in miles per gallon for the aeroplane.

The result is an average figure, since there are different requirements at take-off, compared with cruising at altitude.

On the same journey, approximately 20 kg of nitrogen oxides could be emitted by the aeroplane.

g  How do nitrogen oxides form in an aircraft engine? Write an equation for the formation of nitrogen monoxide, NO.

h  Assume that the 20 kg consists solely of nitrogen monoxide. Calculate the number of moles formed.

i  How many molecules are there in 20 kg of nitrogen monoxide?

It is estimated that a single molecule of nitrogen monoxide can, on average, destroy 1000 molecules of ozone.

j  How many molecules of ozone could be destroyed by 20 kg of nitrogen monoxide?

# Unit **3**

# Practical skills in chemistry 1

# Practical skills

## Chapter 13

During the AS course, you perform experiments that support the theoretical chemistry you are studying. The quality of the results obtained depends partly on the skill with which you manipulate the apparatus. However, the equipment itself also limits the accuracy of quantitative work. This affects the results obtained and, therefore, the reliability of the conclusions drawn.

The level of uncertainty in an experiment can be indicated when recording results. In a simple way, it is addressed by using the appropriate number of significant figures.

An additional source of error in an experiment is weakness in the procedure. Such errors are hard to quantify but should be considered in an overall assessment of the reliability of experiment.

This chapter considers issues relevant to the tasks that will be set to assess your practical skills.

## Accuracy of measurements and significant figures

Each piece of laboratory apparatus used to measure a quantity has a limit of accuracy. For example, a basic balance may give a measurement to the nearest gram; greater accuracy can be achieved by using a balance capable of reading to one, two, three or even four decimal places. The volume of a liquid can be measured using a graduated beaker, a measuring cylinder or volumetric apparatus. Each of these has a different level of accuracy. For most measuring equipment, the manufacturer gives the limits of accuracy and these may be etched onto the apparatus. Not all experiments require accurate quantities to be measured, but if a quantitative result is needed, you should be aware of the limitations of the apparatus used.

This inaccuracy is inevitable and is distinct from the competence with which the experiment is performed or the care taken when making a measurement.

The words 'reliable', 'precision' and 'accuracy' have specific meanings. The explanations below can be used as a guide:

- A reliable piece of apparatus is one that is suitable for the purpose for which it is intended.
- A reliable result is one that, using the apparatus provided, is reproducible every time the experiment is repeated.

- An accurate result is one that is close to the true result. This may not be achievable with a particular experimental set-up.
- A precise reading is the best that can be obtained using the apparatus provided and which has been read to the greatest accuracy of which the apparatus is capable.

These differences are subtle and become apparent only as you carry out more and more experiments. It is, however, important to consider each of these concepts whenever you do practical work.

# Measurement of mass

If a **balance** measures to one decimal place, the mass is normally accurate to 0.1 g (sometimes 0.05 g). For a balance measuring to two decimal places, the mass is accurate to 0.01 g (sometimes 0.005 g). If carefully maintained and cleaned, a four-decimal-place balance can be accurate to 0.0001 g.

So if 1 g is measured using a balance accurate to two decimal places, the mass will have an inevitable error of 0.01 g. This means that the true mass lies anywhere between 0.99 g and 1.01 g. It is useful to express this as a percentage error because this can then be carried forward and applied to results that are based on this measurement.

The percentage error is:

$$\frac{0.01}{1} \times 100 = 1\%$$

Any result based on this measurement would also have a 1% error. In practice, experiments usually involve more than one measurement and this increases the uncertainty in the final answer.

A 1% error might be acceptable for most experimental work. However, weighing 1 g using a balance accurate to four decimal places would give an error of only

$$\frac{0.0001}{1} \times 100 = 0.01\%$$

If the mass measured is much smaller, more thought needs to be given. A balance accurate to two decimal places used to measure a mass of 0.10 g would have the percentage error:

$$\frac{0.01}{0.10} \times 100 = 10\%$$

This is a significant error. Using a balance accurate to four decimal places to weigh a mass of 0.10 g would give a more acceptable percentage error of 0.10%.

# Measurement of volume

There is a wide choice of equipment available to measure volumes of liquids. **Beakers** and **conical flasks** may be graduated to indicate volumes; **measuring cylinders** vary in size from 10 cm$^3$ to 2000 cm$^3$.

**Volumetric apparatus** includes pipettes, burettes and volumetric flasks. These are more accurate, but their accuracy still has limits.

# Basic equipment

## Beakers and conical flasks

Markings on the sides of beakers and conical flasks serve only as a rough guide and it is impossible to assess the error involved. However, it is likely to be large. Hence, beakers and conical flasks are not suitable for measuring volumes reliably.

## Measuring cylinders

A useful rule of thumb for measuring cylinders is that they are accurate to one-half of the difference between the graduation marks. If a $25\,cm^3$ measuring cylinder has graduations every $0.5\,cm^3$, the error is $0.25\,cm^3$. If it is used to measure a volume of $20\,cm^3$, then the true volume is between $19.75\,cm^3$ and $20.25\,cm^3$. The percentage error is:

$$\frac{0.25}{20.0} \times 100 = 1.25\%$$

A $100\,cm^3$ measuring cylinder may have graduations every $1\,cm^3$. The error is, therefore, $0.5\,cm^3$. So, $20\,cm^3$ measured in a $100\,cm^3$ cylinder is $20 \pm 0.5\,cm^3$. The percentage error is:

$$\frac{0.5}{20.0} \times 100 = 2.5\%$$

A $250\,cm^3$ measuring cylinder usually has graduations every $2\,cm^3$. The error is, therefore, $1\,cm^3$. So, $20\,cm^3$ measured in a $250\,cm^3$ cylinder is $20 \pm 1\,cm^3$. This has a large percentage error of 5%.

The difference in the accuracy of these measurements is considerable. It is important to bear this in mind when planning or interpreting the results of an experiment.

Increased inaccuracy arises if start and end graduations are used to measure a volume. For example, if $20\,cm^3$ of water were measured by filling a $100\,cm^3$ measuring cylinder to the $100\,cm^3$ mark and then pouring off the water until the $80\,cm^3$ mark was reached, both the start and the end gradations are subject to an error of $0.5\,cm^3$. It is possible that the starting volume could be $100.5\,cm^3$ and the end volume $79.5\,cm^3$. The possible error is therefore $1\,cm^3$ in the $20\,cm^3$ measured, which is 5% (although this is probably greater than the experimental reality).

*Note*: In planning an experimental procedure, it would probably be inappropriate to use a balance accurate to four decimal places (0.05% accuracy) and then to measure volume to 5% accuracy.

# Volumetric equipment

Measuring cylinders have limited accuracy, so for more accurate work volumetric apparatus is used:

- A **pipette** is usually selected when a small fixed volume of liquid or solution is required.
- A **burette** is used to provide more variable volumes.

## Using a pipette

When a pipette is used correctly it delivers a precise volume (Figure 13.1).

> **ℯ** Solution is sucked into the pipette, so a pipette filler must always be used.

When a 25.0 cm³ pipette is used correctly, a volume of 25.0 cm³ can be measured to within 0.06 cm³. The percentage error is:

$$\frac{0.06}{25.00} \times 100 = 0.24\%$$

This is a big improvement in accuracy compared with a measuring cylinder.

Some pipettes are graduated to allow a volume less than the full capacity to be dispensed. These are useful but are usually less accurate than a standard pipette.

## Using a burette

When used correctly, a burette is very accurate. A funnel should be used to fill a burette so that solution is not spilled down the outside (Figure 13.2). Care must be taken not to overfill. The tap should, of course, be closed during this procedure. However, before recording a volume, some solution should be run out so that the space below the tap is filled.

When using a burette, the required volume is measured as the difference between final and initial readings. It is usual to read from the bottom of the meniscus but, because two readings are being subtracted, the top of the solution could be used. This is sometimes helpful with coloured solutions.

Burettes usually have graduations every 0.1 cm³ and should, therefore, be accurate to 0.05 cm³. The volume delivered is the difference between two measurements, each of which is accurate to 0.05 cm³, so the volume obtained has an error of 0.1 cm³.

A volume measured by a burette is usually not more than 50 cm³.

When the pipette is filled, the meniscus of the liquid should sit on the volume mark on the neck of the pipette. To do this reliably requires practice.

The solution in the pipette should be allowed to run out freely. When this is done a small amount will remain in the bottom of the pipette. Touch the tip of the pipette on the surface of the liquid that has been run out and then ignore any further solution that remains in the pipette.

*Figure 13.1*
*Using a pipette*

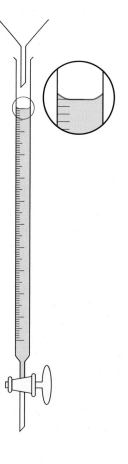

*Figure 13.2*
*A burette*

**ℯ** There are different grades of burette, so it is always worth checking the accuracy level.

### Using a volumetric flask

The mark on the neck of a volumetric flask indicates a specific volume. The accuracy is almost certain to be quoted on the flask and is usually of the order of 0.125%. If a volumetric flask is used to measure 250.0 cm$^3$, the volume obtained has the error:

$$\frac{0.125}{100} \times 250 = 0.313 \, \text{cm}^3$$

This means that the volume is approximately between 249.7 cm$^3$ and 250.3 cm$^3$.

*Note*: If 250 cm$^3$ were measured using a 50 cm$^3$ burette filled five times, the exercise would not only be more tedious, it would also be more inaccurate because the accumulated error would be greater.

A volumetric flask is very accurate. It holds the specified volume when the meniscus of the solution sits on the engraved marking on the neck of the flask.

When dissolving a solid in water to make the solution, care is needed. It is usual to dissolve the solid in a container such as a beaker using a volume of distilled water that is less than that required for the solution. A beaker is used so that, if necessary, the contents can be warmed to encourage the solid to dissolve. Once dissolved, the solution is allowed to cool to room temperature and is then transferred to the volumetric flask, using a funnel. Any solution remaining in the beaker is washed into the flask. Distilled water is added carefully to the flask until the meniscus sits on the engraved mark on the neck. With the stopper firmly in place, the flask is inverted a few times to ensure that the solution is completely mixed.

*e* Don't heat a volumetric flask — it will probably break.

# Measurement of temperature

## Thermometers

As with equipment used to measure volumes, different thermometers have different levels of accuracy. To obtain the percentage error in their readings, it is best to use the graduations. A useful rule of thumb for thermometers is that they are accurate to one-half the difference between their graduation marks.

In some experiments, thermometers are used to measure small differences in temperature. Care is needed in assessing the error resulting from the measurement of temperature change. In order to determine a temperature difference, both the initial and final temperatures are measured, which doubles the error. An error of 0.5°C obtained when using a thermometer graduated in units of 1°C to measure a temperature rise of 2°C is significant, but less so if the rise is 40°C. The percentage errors are 50% and 2.5% respectively.

A digital thermometer should be more accurate than a standard thermometer. It is normal practice with digital equipment to record all the digits shown, but there is still some inherent error.

*e* Don't assume that digital readings are more accurate.

1 Calculate the percentage error in measuring a mass of 2.5 g using a balance accurate to:
   a 0.1 g
   b 0.05 g
   c 0.001 g

2 Calculate the percentage error in measuring 50 cm$^3$ of a solution using:
   a a 250 cm$^3$ measuring cylinder accurate to 1 cm$^3$
   b a 100 cm$^3$ measuring cylinder accurate to 0.5 cm$^3$
   c a 50 cm$^3$ pipette accurate to 0.1 cm$^3$
   d a 25 cm$^3$ pipette accurate to 0.06 cm$^3$, used twice
   e a 50 cm$^3$ burette with each reading accurate to 0.05 cm$^3$

3 Calculate the percentage error in measuring a temperature rise of 6°C using a thermometer accurate to:
   a 0.5°C
   b 0.1°C

4 0.25 g of calcium carbonate is weighed out to an accuracy of 0.05 g and is then reacted with excess hydrochloric acid. The volume of carbon dioxide formed is 60 cm$^3$ and is collected in a syringe accurate to 1 cm$^3$.
   a Why does it matter that the hydrochloric acid is in excess?
   b Calculate whether the balance or the syringe contributes the more significant error in this experiment.
   c Write the equation for the reaction.
   d Assuming that the volume of 1 mol of gas in the experiment is 24 dm$^3$, show that the volume of gas collected would be 60 cm$^3$.
   e If the acid used has a concentration of 0.1 mol dm$^{-3}$, calculate the volume that would be required to ensure that there were twice as many moles of hydrochloric acid as calcium carbonate.

# Recording results

When recording data, the accuracy should be indicated appropriately. For example, if using a balance that reads to two decimal places, the masses recorded should indicate this. This may seem obvious for a mass of, for example, 2.48 g. However, you must remember that this applies equally for a mass of, for example, 2.50 g. Here the '0' should be included after the '5' to indicate that this mass is also accurate to two decimal places. Recording the mass as 2.5 g is incorrect and could be penalised in an exam.

When using a measuring cylinder that is accurate to only 1 cm$^3$ to measure a volume, the result would normally be recorded without any decimal places.

Burette readings are normally recorded to 0.05 cm$^3$ as this represents the limit of accuracy that is possible. So a reading of 23.45 cm$^3$ is acceptable, but 23.47 cm$^3$ is not. The reading should not be recorded as 23.4 cm$^3$ because this implies an accuracy limit of only 0.1 cm$^3$. If the initial reading is zero, it should be written as 0.00 cm$^3$ because this reading is measured to the same level of accuracy as any other.

## Overall accuracy of experimental results

When performing a quantitative experiment it is likely that you will use several pieces of apparatus of varying accuracies. For example, a mass might be accurate to 0.5%, but a temperature might be accurate to only 4%. It is important to consider the accuracy of a value calculated using these data. The effect of inaccuracies in the equipment on a final answer can be involved.

### Example of the effect of inaccuracies

Suppose that in an enthalpy-change experiment, there is a mass of 30.25 g (accurate to two decimal places). The temperature change is 5.0°C, measured with a thermometer accurate to 0.1°C. On handling the results, these figures are multiplied together.

Remember that the starting temperature might have a 0.1°C error and the final temperature might also have a 0.1°C error. Therefore, the temperature change could be inaccurate by as much as 0.2°C. The mass could lie between 30.24 g and 30.26 g.

mass multiplied by temperature = $30.25 \times 5 = 151.25$ g °C

However, consider the possible errors in the readings. The true result could be:

$30.24 \times 4.8 = 145.152$ g °C

or, at the other extreme:

$30.26 \times 5.2 = 157.352$ g °C

The result is, therefore, uncertain. It is apparent that the result of this experiment should not be written as 151.25 g °C without qualifying it in some way.

It would be better to record the result as 151 g °C, which gives some indication that you are aware of the limited accuracy. Alternatively, it could be recorded as 151.25 g °C ±4% to indicate that the temperature change was only accurate to 4%. However, the section on significant figures (below) provides a better way of doing this.

### Rounding up

If a calculation involves several steps, it is only when the final answer is obtained that you should consider the potential inaccuracy. It is a mistake to round a calculation after one step and then use the rounded answer in the next. If you wish to quote each step to an appropriate accuracy then you can do so, but you should always carry the full number of decimal places onto the next step. Otherwise, further errors could be introduced.

## Significant figures

Errors in measurements and results have been discussed, as has the importance of using a suitable number of figures when results are recorded, i.e. the correct number of **significant figures**. Applied properly, this indicates the level of accuracy.

- If the number is a decimal less than 1, the number of significant figures is the number of digits after left-hand zeros beyond the decimal point.

*e* You will not be required to carry out a detailed analysis. However, you should be aware of the effect of the least accurate measurement on the final answer.

◀ The percentage error in measuring the temperature is
$$\frac{0.2}{5} \times 100 = 4\%$$
The percentage error in measuring the mass is
$$\frac{0.01}{30.25} \times 100 = 0.03\%$$

*e* Inaccuracy is inherent in the equipment used and is independent of the precision with which the experiment was performed.

- 0.12 has two digits beyond the decimal point. It has two significant figures and implies that the accurate number is between 0.11 and 0.13. In other words, the *final digit* is considered to have an accuracy of $\pm 1$.
- 0.120 has three digits beyond the decimal point. Therefore, it has three significant figures and the number lies between 0.119 and 0.121.
- 0.004 has just one significant figure, because the two zeros beyond the decimal point are not counted. The number is between 0.003 and 0.005.
- 0.0042 has two significant figures. The number lies between 0.0041 and 0.0043.
- 0.00400 has three significant figures. The two zeros beyond the '4' do count, so the number lies between 0.00399 and 0.00401.

A number such as 151.25 (calculated in the enthalpy example on page 249) has a five-figure accuracy; it has five significant figures. The true number lies between 151.24 and 151.26.

If 151.25 is rounded to 151, then it has three significant figures. This indicates that the number lies between 150 and 152.

Writing 150 does not make the meaning clear. This is because when the final digit is zero it implies that the number is either between 149 and 151 or between 140 and 160. To get around this difficulty, scientific notation is used:
- $1.50 \times 10^2$ indicates that the number lies between 149 and 151 (remember the zero is a significant figure)
- $1.5 \times 10^2$ indicates that the number lies between 140 and 160.

For the enthalpy experiment example (page 249), $1.5 \times 10^2$ is an appropriate indication of the accuracy.

**e** As well as its importance in practical work, examination questions often expect you to give an answer to an appropriate number of significant figures.

## Questions

1 Using the appropriate number of significant figures, write down the mass of exactly 14 g measured using a balance accurate to:
   **a** two decimal places
   **b** four decimal places

2 Using the appropriate number of significant figures, write down the volume of exactly 20 cm$^3$ measured using:
   **a** a measuring cylinder accurate to 1 cm$^3$
   **b** a pipette accurate to 0.06 cm$^3$

3 A student performed a titration using a burette accurate to 0.05 cm$^3$. The student's results are tabulated below.

| Titration | Rough | 1 | 2 | 3 |
|---|---|---|---|---|
| Final volume/cm$^3$ | 24.0 | 23.55 | 23.9 | 23.60 |
| Initial volume/cm$^3$ | 0 | 0 | 0 | 0 |
| Volume used/cm$^3$ | 24 | 23.55 | 23.9 | 23.60 |

   **a** Copy this table, but give the results to the correct number of significant figures.
   **b** State the average value for the titre that should be used in any subsequent calculation.

# Errors and improvements in experimental procedure

## Errors in procedure

Apart from the limitations imposed by apparatus, experiments also have errors caused by the procedure. Such errors are difficult to quantify, but the following check list might help you to assess an experiment.

### Handling substances

- Can you be sure that the substances you are using are pure? Solids may be damp and if a damp solid is weighed, the absorbed moisture is included in the mass.
- Does the experiment involve a reactant or a product that could react with the air? Remember that air contains carbon dioxide, which is acidic and reacts with alkalis. Some substances react with the oxygen present. Air is also always damp.

### Heating substances

- If you need to heat a substance until it decomposes, can you be sure that the decomposition is complete?
- Is it possible that heating is too strong and the product has decomposed further?

### Solutions

- If an experiment is quantitative, can you guarantee that any solution used is exactly at the stated concentration?

### Gases

- If a gas is collected during the experiment, can you be sure that none escapes?
- If you collect a gas over water, are you sure it is not soluble?
- The volume of a gas is temperature-dependent.

### Timing

- If an experiment involves timing, are you sure that you can start and stop the timing exactly when required?

### Enthalpy experiments

- Heat loss is always a problem in enthalpy experiments, particularly if the reaction is slow.

## Improving experiments

As with limitations resulting from procedure, it is not possible to provide a comprehensive list of ways to improve experiments. However, it is worth emphasising that simple changes are often effective.

It is a mistake to imagine that perfection can be achieved by employing complicated devices. Using more accurate measuring equipment is only useful if it focuses on the significant weakness. For example, if the volume of a solution is not required accurately, then there is no point in using volumetric equipment.

If a change in the apparatus seems necessary, look for a straightforward alternative — for example:

- If the problem is that a gas that is to be collected over water is slightly soluble, using a syringe might be an effective remedy.
- If a gas might escape from the apparatus when a reagent is added, the procedure could be improved by placing one reagent in a container inside the flask containing the other reagent and mixing them when the apparatus is sealed.
- If the problem with an enthalpy reaction is that it is too slow, using powders would help.

# Summary

You should now be able to:

- understand the limitations of accuracy of equipment to measure mass, volume and temperature
- calculate the percentage error involved in making measurements
- understand what is meant by the use of significant figures
- understand the importance of recording results to an appropriate number of significant figures
- consider the reliability of an experimental procedure and suggest simple improvements

It is straightforward to decide on the error of a single measurement using a single piece of apparatus, but it can be more complicated if several measurements are made and the results are used to calculate a further quantity.

In the simplest situation, two readings, $x$ and $y$, might be added or subtracted to provide a further figure, $(x + y)$ or $(x − y)$. Suppose there is an error of $\delta x$ in measurement $x$ and an error of $\delta y$ in measurement $y$. Then the rule is that the error in $(x + y)$ or $(x − y)$ will be given by $\sqrt{(\delta x)^2 + (\delta y)^2}$. The error is the same whether the figure is added or subtracted.

A burette measures the difference between two volumes. If each reading is subject to a possible error of $0.05\,cm^3$, this can potentially lead to an error of $0.1\,cm^3$ if the first reading is over-estimated by $0.05\,cm^3$ and the second is underestimated by $0.05\,cm^3$. However, this is actually very unlikely to occur and the rule above gives the more reasonable estimation of the uncertainty in the volume measured as $\sqrt{(0.05)^2 + (0.05)^2} = \sqrt{(0.0025)^2 + (0.0025)^2} = 0.07\,cm^3$.

The measurements do not have to be from the same piece of apparatus for the rule to apply. If $5.00\,cm^3$ measured using a pipette accurate to $0.06\,cm^3$ was added to $50\,cm^3$ measured using a measuring cylinder accurate to $0.5\,cm^3$, then the uncertainty in the $55\,cm^3$ obtained would be $\sqrt{(0.06)^2 + (0.5)^2} = \sqrt{(0.0036)^2 + (0.25)^2} = 0.504\,cm^3$. Notice the rather obvious point that it is the measuring cylinder that provides most of the uncertainty in the final volume.

Quite often when processing results, a reading is multiplied by a factor. In this case, any error in the reading is simply magnified by the multiplying factor. Suppose that, in an experiment, $1.235\,g$ of copper carbonate (which for this illustration is assumed to be completely accurate) when heated produced $236\,cm^3$ of carbon dioxide measured in a syringe accurate to within $2\,cm^3$. Then it would be deduced that the volume of carbon dioxide produced from a mole of copper carbonate (molar mass $123.5\,g\,mol^{-1}$) is $236 \times 100 = 23600\,cm^3$ and the uncertainty is magnified by the same amount, i.e. $2 \times 100 = 200\,cm^3$.

Other experiments might involve multiplying or dividing quantities. In this case, the calculation of the uncertainty becomes rather more complicated. In fact, if an answer, $A$, is calculated by multiplying $X$ with an error $\delta X$, by $Y$, with an error $\delta Y$, then the error in the answer, $\delta A$, is obtained from the expression:

$$\frac{\delta A}{A} = \sqrt{\left(\frac{\delta X}{X}\right)^2 + \left(\frac{\delta Y}{Y}\right)^2}$$

The same formula is used if $X$ and $Y$ are divided. For example, in an enthalpy calculation, the heat produced in a reaction is dispersed over a volume of $50\,g$ (accurate to $1\,g$) and a temperature rise of $3.5°C$ (accurate to $0.7°C$) is observed. Then the uncertainty in the result ($\delta A$) obtained by multiplying the mass by the temperature rise is calculated as follows:

mass × temperature rise, $A$, $= 50 \times 3.5 = 175\,g°C$

$$\frac{\delta A}{175} = \sqrt{\left(\frac{1}{50}\right)^2 + \left(\frac{0.7}{3.5}\right)^2} = \sqrt{(0.0004 + 0.04)} = 0.201$$

$\delta A = 0.201 \times 175 = 35.2\,g°C$

So the figure of 175 has an uncertainty of 35.2. This is quite substantial, largely because of the high percentage error in the temperature measurement.

Often, tedious calculations can be avoided by a little thought. In this case, the error caused by the thermometer is 20% (0.7/3.5) and this is much more significant than the error in the volume of 2% (1/50), so it could be anticipated that the error in the answer to the multiplication would be close to 20%. 20% of 175 is 35, which reinforces this point. However, the method used above might be required if the errors in the equipment are more equivalent.

A complete analysis of the uncertainty of an experiment requiring the measurement of enthalpies calculated from $mc\Delta t$ and then the application of Hess's law would be a fairly daunting prospect. The mass would have an error resulting from the balance, the thermometer would record a temperature change involving the subtraction of two readings and the conversion to molar quantities would magnify the errors. Finally, when Hess's law is used, the errors in each value in the cycle would have to be taken into account.

## Extension question: iron and sulphuric acid

Iron and sulphuric acid react according to the equation:

$$Fe(s) + H_2SO_4(aq) \rightarrow FeSO_4(aq) + H_2(g)$$

A reaction is performed using:

- 0.70g of iron weighed on a balance with an overall error of 0.01g
- 25cm³ of 2.00mol dm⁻³ sulphuric acid measured using a measuring cylinder accurate to 0.5cm³

As a result, 290cm³ of hydrogen was collected.

a Calculate the percentage error in the measurements of the mass of iron and the volume of sulphuric acid.

b Calculate the number of moles of iron used.

c Calculate the number of moles of sulphuric acid used.

d Consider your answers to parts b and c. Determine the overall measurement error in the experiment.

e (i) Show that the volume of hydrogen that should be collected is 300cm³. (Assume that the volume is at room temperature and pressure, so the volume of 1mol is 24dm³.)

(ii) Can the overall error in the experiment account for the fact that only 290cm³ of gas is collected? Explain your answer.

(iii) A student suggests that an important issue could be that the temperature of the gas produced in the reaction is higher than room temperature. Comment on this suggestion.

(iv) Another suggestion is that the volume is less because the reaction is slow. Comment on this suggestion.

(v) A further suggestion is that gas might escape from the apparatus when the iron is added. Describe how the experiment could be set up to avoid this problem. Draw a diagram of suitable apparatus.

# Index

Group

# The periodic table

**Key:**

| Relative atomic mass |
| **Atomic symbol** |
| name |
| Atomic (proton) number |

| Period | 1 | 2 | | | | | | | | | | | | 3 | 4 | 5 | 6 | 7 | 0 |
|---|---|---|---|---|---|---|---|---|---|---|---|---|---|---|---|---|---|---|---|
| 1 | 1.0 **H** hydrogen 1 | | | | | | | | | | | | | | | | | | 4.0 **He** helium 2 |
| 2 | 6.9 **Li** lithium 3 | 9.0 **Be** beryllium 4 | | | | | | | | | | | | 10.8 **B** boron 5 | 12.0 **C** carbon 6 | 14.0 **N** nitrogen 7 | 16.0 **O** oxygen 8 | 19.0 **F** fluorine 9 | 20.2 **Ne** neon 10 |
| 3 | 23.0 **Na** sodium 11 | 24.3 **Mg** magnesium 12 | | | | | | | | | | | | 27.0 **Al** aluminium 13 | 28.1 **Si** silicon 14 | 31.0 **P** phosphorus 15 | 32.1 **S** sulphur 16 | 35.5 **Cl** chlorine 17 | 39.9 **Ar** argon 18 |
| 4 | 39.1 **K** potassium 19 | 40.1 **Ca** calcium 20 | 45.0 **Sc** scandium 21 | 47.9 **Ti** titanium 22 | 50.9 **V** vanadium 23 | 52.0 **Cr** chromium 24 | 54.9 **Mn** manganese 25 | 55.8 **Fe** iron 26 | 58.9 **Co** cobalt 27 | 58.7 **Ni** nickel 28 | 63.5 **Cu** copper 29 | 65.4 **Zn** zinc 30 | | 69.7 **Ga** gallium 31 | 72.6 **Ge** germanium 32 | 74.9 **As** arsenic 33 | 79.0 **Se** selenium 34 | 79.9 **Br** bromine 35 | 83.8 **Kr** krypton 36 |
| 5 | 85.5 **Rb** rubidium 37 | 87.6 **Sr** strontium 38 | 88.9 **Y** yttrium 39 | 91.2 **Zr** zirconium 40 | 92.9 **Nb** niobium 41 | 95.9 **Mo** molybdenum 42 | [98] **Tc** technetium 43 | 101.1 **Ru** ruthenium 44 | 102.9 **Rh** rhodium 45 | 106.4 **Pd** palladium 46 | 107.9 **Ag** silver 47 | 112.4 **Cd** cadmium 48 | | 114.8 **In** indium 49 | 118.7 **Sn** tin 50 | 121.8 **Sb** antimony 51 | 127.6 **Te** tellurium 52 | 126.9 **I** iodine 53 | 131.3 **Xe** xenon 54 |
| 6 | 132.9 **Cs** caesium 55 | 137.3 **Ba** barium 56 | 138.9 **La** lanthanum 57 | 178.5 **Hf** hafnium 72 | 180.9 **Ta** tantalum 73 | 183.8 **W** tungsten 74 | 186.2 **Re** rhenium 75 | 190.2 **Os** osmium 76 | 192.2 **Ir** iridium 77 | 195.1 **Pt** platinum 78 | 197.0 **Au** gold 79 | 200.6 **Hg** mercury 80 | | 204.4 **Tl** thallium 81 | 207.2 **Pb** lead 82 | 209.0 **Bi** bismuth 83 | [209] **Po** polonium 84 | [210] **At** astatine 85 | [222] **Rn** radon 86 |
| 7 | [223] **Fr** francium 87 | [226] **Ra** radium 88 | [227] **Ac** actinium 89 | [261] **Rf** rutherfordium 104 | [262] **Db** dubnium 105 | [266] **Sg** seaborgium 106 | [264] **Bh** bohrium 107 | [277] **Hs** hassium 108 | [268] **Mt** meitnerium 109 | [271] **Ds** darmstadtium 110 | [272] **Rg** roentgenium 111 | | | | | | | | |

Elements with atomic numbers 112–116 have been reported but not fully authenticated

| 140.1 **Ce** cerium 58 | 140.9 **Pr** praseodymium 59 | 144.2 **Nd** neodymium 60 | 144.9 **Pm** promethium 61 | 150.4 **Sm** samarium 62 | 152.0 **Eu** europium 63 | 157.2 **Gd** gadolinium 64 | 158.9 **Tb** terbium 65 | 162.5 **Dy** dysprosium 66 | 164.9 **Ho** holmium 67 | 167.3 **Er** erbium 68 | 168.9 **Tm** thulium 69 | 173.0 **Yb** ytterbium 70 | 175.0 **Lu** lutetium 71 |
|---|---|---|---|---|---|---|---|---|---|---|---|---|---|
| 232 **Th** thorium 90 | [231] **Pa** protactinium 91 | 238.1 **U** uranium 92 | [237] **Np** neptunium 93 | [242] **Pu** plutonium 94 | [243] **Am** americium 95 | [247] **Cm** curium 96 | [245] **Bk** berkelium 97 | [251] **Cf** californium 98 | [254] **Es** einsteinium 99 | [253] **Fm** fermium 100 | [256] **Md** mendelevium 101 | [254] **No** nodelium 102 | [257] **Lr** lawrencium 103 |